Norbert Peters
Jürgen Warnatz (Eds.)

Numerical Methods in Laminar Flame Propagation

Notes on Numerical Fluid Mechanics
(Series Editor: Karl Förster, Stuttgart)

Volume 6

Volume 1 Boundary Algorithms for Multidimensional Inviscid Hyperbolic Flows (Karl Förster, Ed.)

Volume 2 Proceedings of the Third GAMM-Conference on Numerical Methods in Fluid Mechanics (Ernst Heinrich Hirschel, Ed.)

Volume 3 Numerical Methods for the Computation of Inviscid Transonic Flows with Shock Waves (Arthur Rizzi/Henri Viviand, Eds.)

Volume 4 Shear Flow in Surface-Oriented Coordinates (Ernst Heinrich Hirschel/Wilhelm Kordulla)

Volume 5 Proceedings of the Fourth GAMM-Conference on Numerical Methods in Fluid Mechanics (Henri Viviand, Ed.)

Volume 6 Numerical Methods in Laminar Flame Propagation (Norbert Peters/Jürgen Warnatz, Eds.)

Manuscripts should be well over 100 pages. As they will be reproduced fotomechanically they should be typed with utmost care on special stationary which will be supplied on request. In print, the size will be reduced linearly to approximately 75 %. Figures and diagrams should be lettered accordingly so as to produce letters not smaller than 2 mm in print. The same is valid for handwritten formulae. Manuscripts (in English) or proposals should be sent to the new series editor: Prof. Dr. E. H. Hirschel, MBB-UFE 122, Postfach 80 11 60, D-8000 München 80.

Norbert Peters/Jürgen Warnatz (Eds)

Numerical Methods in Laminar Flame Propagation

A GAMM-Workshop

With 66 Figures

Springer Fachmedien Wiesbaden GmbH

CIP-Kurztitelaufnahme der Deutschen Bibliothek

Numerical methods in laminar flame propagation:
a GAMM workshop/Norbert Peters; Jürgen Warnatz
(eds.).
(Notes on numerical fluid mechanics; Vol. 6)
ISBN 978-3-528-08080-8 ISBN 978-3-663-14006-1 (eBook)
DOI 10.1007/978-3-663-14006-1

NE: Peters, Norbert [Hrsg.]; Gesellschaft für
Angewandte Mathematik und Mechanik; GT

Originally published by Friedr. Vieweg & Sohn Verlagsgesellschaft mbH, Braunschweig in 1982

ISBN 978-3-528-08080-8

Preface

This volume collects the results of a workshop held at Aachen, West-Germany, Oct. 12 - Oct. 14, 1981. The purpose in bringing together scientists actively working in the field of numerical methods in flame propagation was two-fold:

1. To confront them with recent results obtained by large activation-energy asymptotics and to check these numerically.
2. To compare different numerical codes and different transport models for flat flame calculations with complex chemistry.

Two test problems were formulated by the editors to meet these objectives. Test problem A was an unsteady propagating flat flame with one-step chemistry and Lewis number different from unity while test problem B was the steady, stoichiometric hydrogen-air flame with prescribed complex chemistry. The participants were asked to solve one or both test problems and to present recent work of their own choice at the meeting.

The results of the numerical calculations of test problem A are challenging just as much for scientists employing numerical methods as for those devoted to large activation-energy asymptotics: Satisfactory agreement between the five different groups were obtained only for two out of six cases, those with Lewis number Le equal to one. The very strong oscillations that occur at $Le = 2$ and a nondimensional activation energy of 20 were accurately resolved only by one group. This case is particular interesting because the asymptotic theory so far predicts instability but not oscillations. This stimulated a numerical study by B. Rogg, who determined the boundary between the stable and oscillating regimes. Although the paper was not presented at the meeting, it is included in this volume.

On the other hand, test problem B confirmed the expectation that the existing codes for one-dimensional flat flames with complex chemistry yield satisfactory results if a non-uniform grid is used with enough (about 25) grid points within the flame front. With the kinetics prescribed, the fairly large deviations in calculated flame speeds are due to the different transport models used. In spite of the computational effort involved it seems to be neccessary to use rather complex transport models. Some of the papers of Part II of this volume deal with this question.

The workshop was sponsored by the Deutsche Forschungsgemeinschaft, to which the organiser and the participants owe a special debt. We also note with thanks the contribution of Ms. B. Tries in the careful preparation of a major part of the typescripts.

Aachen, April 1982

Norbert Peters

Jürgen Warnatz

Contents

DISCUSSION OF TEST PROBLEM A

by

N. Peters

Institut für Allgemeine Mechanik
Rheinisch-Westfälische Technische Hochschule Aachen
Aachen, West Germany

ABSTRACT

This is a summary of the results of the five groups that have calculated the test problem A of the GAMM-Workshop on Numerical Methods in Laminar Flame Propagation. The test problem is formulated and the six different test cases are defined. The numerical methods of solution and the used grid point system are described and the numerical results are discussed. For the cases with $Le = 1$ a good agreement is achieved with all methods. For $Le > 2$ oscillations may occur that require a very sensitive grid adaption in time direction. It is concluded that extreme care must be taken in calculating unsteady flames with Lewis-numbers different from unity.

1. INTRODUCTION

The test problem A treats the one-dimensional unsteady flame under the assumption of a one-step irreversible reaction

$$X_1 \longrightarrow X_2 \quad .$$

However, the results are not restricted to a unimolecular reaction. It may be shown, that the presence of the second reactant X_2 in a one-step reaction of the form

$$X_1 + X_2 \longrightarrow X_3 + X_4$$

does not change the reaction rate as long as X_1 is deficient and X_2 is largely in excess. In this case X_2 may be treated approximately as an inert and the concentration of X_2 that may appear in the reaction rate may be kept constant. For fuel-oxidizer systems, for example, the reaction rate of a bimolecular reaction takes the form of the reaction rate used in this test problem, if the equivalence ratio is far from stoichiometry. The Lewis number to be used in this case is then the one of the deficient reactant, fuel in case of a lean mixture, oxidizer in case of a rich mixture.

The approximation under which the above is valid is that of a large nondimensional activation energy. For hydrocarbon reactions the nondimensional activation energy is around $\beta = 10$. As a consequence the approximation is valid for equivalence ratios larger than about 1.1 and smaller than about 0.9.

The treatment of flames in the asymptotic limit of large activation energy has attracted much attention in the recent years (Ludford /1/). Under the assumption that the flame is steady, an analytic expression for the flame velocity can be derived (Bush and Fendell /2/). The stability characteristics of the unsteady problem can be analysed under the assumption that the flame is slowly varying (Sivashinsky /3/, Peters /4/). The asymptotic treatment may therefore guide the discussion of the numerical results. On the other hand, from the numerical calculation of this test problem, and particularly from the extension of the parameter range by Rogg /5/, it becomes evident that some properties of the solution, i.e. the oscillations, demand a more sophisticated asymptotic analysis than has been done so far. Again asymptotic and numerical investigations seem to complement each other.

2. FORMULATION OF TEST PROBLEM A

The governing equations for the one-dimensional unsteady flame may be written as

Continuity:

$$\frac{\partial \rho'}{\partial t'} + \frac{\partial (\rho' v')}{\partial x'} = 0 \tag{1}$$

Temperature:

$$\rho' \frac{\partial T'}{\partial t'} + \rho' v' \frac{\partial T'}{\partial x'} = \frac{\partial}{\partial x'} \left(\frac{\lambda'}{c_p} \frac{\partial T'}{\partial x'}\right) + \frac{(-\Delta h')}{c_p} r' \tag{2}$$

Mass fraction of reactant:

$$\rho' \frac{\partial Y'}{\partial t'} + \rho' v' \frac{\partial Y'}{\partial x'} = \frac{\partial}{\partial x'} \left(\rho' D' \frac{\partial Y'}{\partial x'}\right) - M_1 r' \tag{3}$$

Here ρ' is the density, v' the velocity, t' the time coordinate and x' the space coordinate. In the temperature equation T' is the temperature, λ' the conductivity and $(-\Delta h')$ the (positive) heat of reaction. For simplicity it has been assumed that the heat capacity c_p is constant and equal for all species. Furthermore Y' is the mass fraction of the reactant 1 and D' its diffusion coefficient and M_1' its molecular weight. The reaction rate is defined as

$$r' = B' \frac{\rho' Y_1'}{M_1} \exp\left(- \frac{E'}{R' T'}\right) \quad . \tag{4}$$

Here B' is the frequency factor, E' the activation energy and

R' the universal gas constant.

If the upstream concentration is Y'_{1u} and Y'_1 is completely consumed during the reaction, the associated rise in temperature is

$$T'_b - T'_u = \frac{(-\Delta h')Y'_{1u}M_1}{c_p} . \tag{5}$$

The transport coefficients are related to each other by the Lewis number

$$Le = \frac{\lambda'}{\rho' D' c'_p} \tag{6}$$

which is assumed constant different from one. Using the coordinate transformation

$$\begin{aligned} x &= \frac{c'_{pu} v'_{ref}}{\lambda'_u} \int_0^{x'} \rho' dx' \\ t &= \frac{t' c'_{pu} \rho'_u v'^2_{ref}}{\lambda'_u} \end{aligned} \tag{7}$$

with the transformation rules

$$\begin{aligned} \frac{\partial}{\partial t'} &= \frac{\partial x}{\partial t'} \frac{\partial}{\partial x} + \frac{\partial t}{\partial t'} \frac{\partial}{\partial t} \\ \frac{\partial}{\partial x'} &= \frac{\partial x}{\partial x'} \frac{\partial}{\partial x} \end{aligned} \tag{8}$$

one obtains the nondimensional equations

$$\begin{aligned} \frac{\partial T}{\partial t} &= \frac{\partial^2 T}{\partial x^2} + R \\ \frac{\partial Y}{\partial t} &= \frac{1}{Le} \frac{\partial^2 Y}{\partial x^2} - R \end{aligned} \tag{9}$$

independent of the continuity equation. The nondimensional temperature and the reduced concentration are defined

$$T = \frac{T' - T'_u}{T'_b - T'_u} , \qquad Y = \frac{Y'_1}{Y'_{1u}} \tag{10}$$

and the nondimensional reaction rate is

$$R = \frac{\lambda_u'}{c_{pu}' \rho_u' v_{ref}'^2} \frac{r}{\rho} = B\, Y \exp\left(- \frac{\beta(1-T)}{1-\alpha(1-T)}\right) \tag{11}$$

where

$$B = \frac{B' \lambda_u'}{c_{pu}' \rho_u' v_{ref}'^2} \quad . \tag{12}$$

Additional parameters of the problem are the nondimensional activation energy

$$\beta = \frac{E'(T_b' - T_u')}{R' T_b'^2} \tag{13}$$

and the nondimensional heat release parameter

$$\alpha = \frac{T_b' - T_u'}{T_b'} \quad . \tag{14}$$

For simplicity it has been assumed that $\rho'\lambda'/c_p' = (\rho'\lambda'/c_p')_u$.

In eqs. (7) and (12) the index u denotes properties evaluated at upstream conditions and v_{ref}' the reference flame velocity. This reference value is the one that would be obtained for the steady problem in the limit of an infinitely large activation energy. Then, using an asymptotical analysis, one may show that

$$B = \frac{\beta^2}{2\,\mathrm{Le}} \quad . \tag{15}$$

The asymptotic and numerical calculations will establish deviations from this limit for finite activation energies. For the system of coupled parabolic equations defined by eqs. (9), (11) and (15) the following initial and boundary conditions are prescribed

Initial conditions: $t = 0$

$$x \leq 0 \qquad T = \exp x, \qquad Y = 1 - \exp(\mathrm{Le}\, x) \tag{16}$$

$$x \geq 0 \qquad T = 1 \qquad (17)$$
$$Y = 0$$

Boundary conditions, $t > 0$:

$$x \to -\infty: \quad T = 0 \; , \quad Y = 1 \qquad (18)$$

$$x \to +\infty: \quad \frac{\partial T}{\partial x} = 0 \; , \quad \frac{\partial Y}{\partial x} = 0 \qquad (19)$$

The parameters to be used are

$$\alpha = 0.8 \; , \quad \beta = 10, 20 \; , \quad Le = 2, 1, 0.5$$

This leads to six cases to be calculated in test problem A which will be refered to in the following way:

β \ Le	2	1	0.5
10	case 1	case 2	case 3
20	case 4	case 5	case 6

The asymptotic treatment up to second order by Bush and Fendell /2/ leads to

$$\frac{2BLe}{\beta^2} = 1 + 2(3\alpha - 2.344 + Le)/\beta + O(\beta^{-2}) \quad . \qquad (20)$$

In terms of the flame velocity, which is proportional to $B^{-1/2}$ according to eq. (12), this leads to the ratio for the actual flame velocity to the reference velocity

$$M = \frac{v_F'}{v_{ref}'} = 1 - (3\alpha - 2.344 + Le)/\beta + O(\beta^{-2}) \quad . \qquad (21)$$

3. THE PARTICIPATING GROUPS

The following 5 groups participated in the solution of the test problem A:

Group 1: T.P. Coffee, J.M. Heimerl
DRDAR - BLI
Ballistic Research Laboratory
Aberdeen Proving Ground, MD 21005, U.S.A.

Group 2: L. Grundl, F. Joos, D. Vortmeyer
Fakultät für Maschinenwesen
Lehrstuhl B für Thermodynamik
Technische Universität München
Arcisstr. 21
8000 München 2, West Germany

Group 3: K. Raith, W. Schönauer
Rechenzentrum der Universität Karlsruhe
Zirkel 2
7500 Karlsruhe, West Germany

Group 4: R.D. Reitz
Department of Mechanical and Aerospace Engineering
Princeton University
Princeton, NJ 08544, U.S.A.

Group 5: P. Faber, H.-J. Thies, N. Peters
Institut für Allgemeine Mechanik
RWTH Aachen
Templergraben 55
5100 Aachen, West Germany

4. DESCRIPTION OF THE NUMERICAL METHODS USED BY THE DIFFERENT GROUPS

There were four different numerical methods used to solve test problem A. Group 1 and Group 2 used the method of lines, specially the package PDECOL, developed by Madsen and Sincovec /5/. It uses a finite element collocation method based on B-Splines /6/ to account for the spatial dependence of the solution

$$T = \sum_{i=1}^{N} T^{(i)}(t)\, B_i(x)$$

$$Y = \sum_{i=1}^{N} Y^{(i)}(t)\, B_i(x) \qquad (22)$$

Where the basis functions $B_i(x)$ are B-Splines. The time-dependent coefficients $T^{(i)}$ and $Y^{(i)}$ are determined by requiring that the expansions (22) satisfy the boundary conditions and the differential equations at N-2 collocation points. The $B_i(x)$ are piecewise polynomials of order KORD-1 which are joined to each other at NB breakpoints in x direction, where NCC continuity conditions are applied. The choice of the breakpoints, the polynomial order and the number of continuity conditions $(1 < NCC < KORD)$ are left to the user of the package. Group 1 used cubic splines (KORD = 4) and NCC = 2 while Group 2 varied KORD from 3 to 11 and NCC from 2 to 7. Both groups used equally spaced breakpoints in the center of the interval of integration, but while Group 1 fixed about half of the breakpoints at increasing step size outside of the center region, Group 2 had 110 out of 115 in the equidistant region. The number of breakpoints of Group 1 varied between 40 (cases 1, 2 and 5), 60 (cases 3 and 6) and 80 (case 4). The length of the interval of integration varied between 20 and 40 for Group 1 and was 400 for Group 2. A further difference in the two methods was that Group 1 had transformed the governing equations prior to calculation while Group 2 used the original equations. A more detailed description of the work done by Group 1 is found in /8/.

Group 3 has developed an own package with a selfadaptive method for the solution of parabolic and elliptic equations /9/, /10/. The method uses an implicit finite difference discretization on a non-equidistant grid. The order of the discretization error is adapted such that a prescribed global error of the solution is not exceeded. For the present problem the x-grid and the x-order q were prescribed $(q = 4)$ while the t-grid and the t-order p was selfadapted $(1 \leq p \leq 5$, mostly $p = 5)$. An estimate of the global error is computed together with the solution. A total number of 121 nonequally spaced or 201 equally spaced grid points was used in x-direction. A coordinate transformation $\bar{x} = x + t$, $\bar{t} = t$ was introduced to keep the flame in the domain of integration $-10 \leq \bar{x} \leq 14$ as long as possible.

Group 4 used a standard explicit finite difference scheme with 101 equally spaced grid points in the region $-20 \leq \bar{x} \leq 5$ (cf. /11/). The method was first order in time and second order in space. No transformation of the coordinate system was performed prior to the numerical calculation.

Group 5 used an implicit finite difference method in a transformed coordinate space $\bar{x} = \exp x$, $\bar{t} = t$. The transformation confines the region $-\infty \leq \bar{x} \leq 0$ into $0 \leq \bar{x} \leq 1$ and leads to a linear initial temperature profile. The method was first order in time and second order in space with an equidistant spacial grid point distribution in the region $0 \leq \bar{x} \leq 4$. The number of grid points was varied between 40 and 400.

5. RESULTS

In the reply form following the announcement of the workshop, essentially two results were requested:

1.) The flame velocity $-\partial x_F/\partial t$ for all six cases together with a comment on whether oscillations were observed
2.) The enthalpy $T + Y - 1$ at $x = +\infty$

Table 1

FLAME VELOCITIES

case / Group	1 Le= 2 β=10	2 Le= 1 β=10	3 Le= 0.5 β=10	4 Le= 2 β=20	5 Le= 1.0 β=20	6 Le= 0.5 β=20
1	0.849	0.918	0.959	Oscill.	0.954	0.978
2	0.708 ± 0.11	0.894 ± 0.18	0.968 ± 0.21	0.569 ± 0.115 ÷0.120	0.903 ± 0.21	0.968 ± 0.23
3	0.88	0.91	0.97	Oscill.	not comp.	not comp.
4	0.867 ± 0.005	0.907 ± 0.005	0.956 ± 0.004	0.975 ± 0.105	0.961 ± 0.159	0.975 ± 0.105
5	0.842	0.915	1.1	Oscill.	0.9515	1.2
Asympt. results						
eq.(20)	0.8417	0.908	0.9486	0.9107	0.951	0.9733
eq.(21)	0.7944	0.8944	0.944	0.8972	0.9472	0.9722

The second result was requested in order to obtain a first control about the behaviour of the solution in the burnt gas region and to check whether the domain of integration had been extended sufficiently far downstream. The latter was the case for all groups for the cases 2 and 5 (Le = 1) and for the entire work of Group 1, 2 and 3. Group 4 reported the following deviations from the expected value of zero: 0.0003 for case 1, 0.0016 for case 3, 0.0014 for case 4 and 0.0015 for case 6. Group 5 observed 0.005 in case 1 and 0.03 and 0.02 in the cases 3 and 6, respectively.

The results for the flame velocity are summarized in Table 1 and compared to the asymptotic results obtained from eqs. (20) and (21). Both expressions are given to indicate the magnitude of the error associated with the asymptotic formulas, which can expected to be close to the difference of the two values. It is clear from the table that the cases 2 and 5 (Le = 1) were fairly accurately calculated by all groups. These cases do not present any difficulties. This is due to the fact that the two differential equations (9) reduce to a single one for the variable $T = 1 - Y$, which is more easily computed. From the asymptotic analysis it is also clear that this single parabolic differential equation will lead to a stationary solution for the flame velocity. Stationary values of the flame velocity were also computed for the cases 1, 3 and 6 by the Groups 1, 3 and 5. The Groups 2 and 4 obtained some oscillations which they believe are due to the numerical method (Group 4 expresses doubts about this for the cases 5 and 6). The reported oscillations were smaller for the cases with $\beta = 10$ than for those with $\beta = 20$.

It is clear that the accuracy of the computed flame velocity, stationary or not, depends not only on the numerical scheme, but also upon the computational effort, i.e. computer time and storage. Group 5 investigated the effect of the (equidistant) step size and found the following deviations from the asymptotic value calculated by eq. (20):

case \ $\Delta\bar{x}$	0.1	0.033	0.01
1	24.7 %	2.7 %	0.1 %
3	41 %	25 %	16 %
6	51 %	34 %	21 %

The partially large deviations for Le = 0.5 , cases 3 and 6, are not observed by other groups and are probably due to a unsufficient domain of integration.

Oscillations as a property of the solution were observed by the Groups 1, 3 and 5 for the case 4. As an example the result of Group 1 is shown in Fig. 1. Instabilities were to be expected for large activation energies and Le > 1 due to the analysis in /3/, /4/. The more detailed numerical calculation by Rogg /5/ shows that oscillations are damped for Lewis number close to one and relatively small activation energies. According to his Fig. 1 the case 4 just falls beyond the line separating the oscillating from the non-oscillating solutions, while the case 1 falls into the non-oscillating regime.

The very accurate solution by Group 1 shows non-harmonic oscillations with very steep peaks. Minimum flame speeds occur around 0.4 and maximum flame speeds around 6.5. The solutions seem to evolve into a steady limit cycle with a period around t = 9.0. The same period of t = 9.0 , but different amplitudes, were also calculated by Group 3 who carried the calculations only over two periods. Group 5 observed strong oscillations which, however, did not seem accurately enough computed to determine any further property of the solution. There was agreement among all groups that the oscillations in case 4 are a property of the solution.

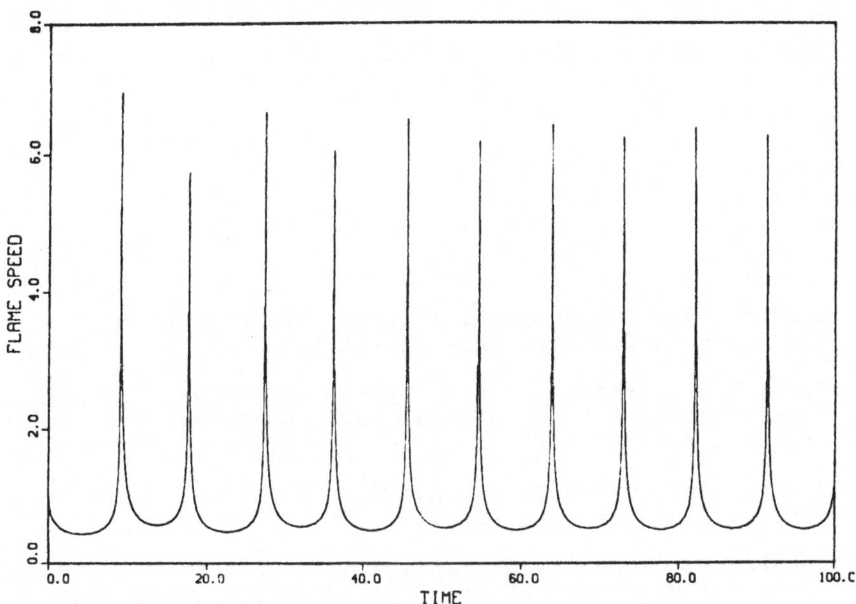

Figure 1: Flame speed for case 4 by Group 1.

6. CONCLUSION

The six cases calculated by the five different groups exhibit different numerical difficulties. The cases with $Le = 1$ are relatively simple since they result in a reduction of the number of differential equations from two to one. The cases with $Le = 0.5$ have stationary flame velocities but they demand special care concerning a sufficiently large domain of integration. For all cases except case 4 satisfactory but not necessarily accurate solutions were calculated by all groups. The particular difficulty associated with the oscillations in case 4 was solved only by Group 1 and Group 3. Group 5 observed strong oscillations but did not obtain any satisfactory results in this case. The fact that only a careful treatment will even detect the oscillation property of the problem is certainly a warning for all future calculations of unsteady flames. For multidimensional calculations it seems necessary to employ moving grid techniques.

REFERENCES

/1/ Ludford, G.S.S.: Activation Energy Asymptotics of the Plane Premixed Flame, This workshop.

/2/ Bush, W.B., Fendell, F.E.: Asymptotic Analysis of Laminar Flame Propagation for General Lewis Numbers, Comb. Sci. Techn. 1 (1970), 421-428.

/3/ Sivashinsky, G.I.: On a Converging Spherical Flame Front, Int. J. Heat Mass Transf. 17 (1974), 1499-1506.

/4/ Peters, N.: Theoretical Implications of Nonequal Diffusivities of Heat and Matter on the Stability of a Plane Premixed Flame, This workshop.

/5/ Rogg, B.: The Effect of Lewis Number Greater than Unity on an Unsteady Propagating Flame with One-Step Chemistry, This workshop.

/6/ Madsen, N.K., Sincovec, R.F.: PDECOL: General Collocation Software for Partial Differential Equations, Preprint UCRL-78263 (Rev 1), Lawrence Livermore Laboratory (1977).

/7/ de Boor, C.: Package for Calculating with B-Splines, SIAM J. Numer. Anal. 14 (1977), 441-472.

/8/ Coffee, T.P.: Flat Flame Olympics: Test Problem A, Internal Report, Ballistic Research Laboratory, Aberdeen Proving Ground (1981).

/9/ Schönauer, W., Raith, K., Glotz, G.: The Principle of the Difference of Difference Quotients as a Key to the Self-adaptive Solution of Non-Linear Partial Differential Equations, Comp. Meth. Apll. Mech. Engng. 28 (1981), 327-359.

/10/ Schönauer, W., Raith, K., Glotz, G.: The SLDGL-Program Package for the Selfadaptive Solution of Nonlinear Systems of Elliptic and Parabolic PDE's, Advances in Computer Methods for Partial Differential Equations - IV, Eds.: R. Vichnevetzky, R.S. Stepleman, IMACS 1981, 117-125.

/11/ Reitz, R.D.: Computations of Laminar Flame Propagation Using an Explicit Numerical Method, 18th Symp. (Int.) on Combust., The Combustion Institute, Pittsburgh 1981, 433-442.

Activation-Energy Asymptotics of the Plane Premixed Flame

G.S.S. Ludford
Theoretical & Applied Mechanics
Cornell University
Ithaca, NY 14853, USA

Abstract. The plane premixed flame is considered in the limit of large activation energy for various chemical reactions. A full account of the one-reactant model is given, including its equivalence to a two-reactant model away from stoichiometry. The behaviour of the two-reactant model close to stoichiometry is then sketched. A brief discussion is given of more complex kinetic schemes, namely: the two-reactant model with dissociation of the product; sequential decomposition reactions; and the autocatalytic chain reaction introduced by Zeldovich. Emphasis throughout is on explaining the method of activation-energy asymptotics and what it offers in the computation of plane flames.

Acknowledgements. This work was supported by the U.S. Army Research Office and by the Alexander von Humboldt Foundation (through a Senior US-Scientist Award).

1. Introduction

Williams has maintained /1/ that activation-energy asymptotics makes all competing approximations to the burning rate of the simplest plane premixed flame obsolete;an idea of the effort expended in approximating the corresponding "laminar-flame eigenvalue" can be obtained by reading chapter 5 of his book /2/. We shall use this simplest problem to explain the method, and then briefly discuss some more complicated kinetic schemes to which it has been applied, all within the context of the steady plane flame.

The results obtained by activation-energy asymptotics are of computational value on three counts. (i) They provide a good starting point, at least for large activation energy; an example is the initial condition given in problem A. (ii) When a computation runs into trouble, they may offer an explanation that makes a search for defects in the numerical scheme unnecessary; as Peters notes /3/, the lack of convergence for $\beta = 20$, $Le = 2$ in problem A can be attributed to an instability predicted by activation-energy asymptotics. (iii) Results for simplified forms of complicated kinetic schemes can identify the essential features; problem B may soon benefit in this way from activation-energy asymptotics. The treatment of more complex kinetics is one of the current thrusts of the method; the prospects include clarifying the role of chain processes in flames and better understanding of catalysis and chemical inhibition /2/. Items (i) and (iii) will be mentioned again at appropriate places.

There is now a book, by Buckmaster and Ludford /4/, on the treatment of laminar flames by means of activation-energy asymptotics. The basic idea is simple and, in a crude form, has been around for a long time as the so-called flame-sheet approximation. Consider the simplest Arrhenius reaction term

$$\Omega = \mathfrak{D} Y \exp(-\theta/T) \tag{1}$$

and let $\theta \to \infty$. If $\mathfrak{D}$ remains bounded, then $\Omega \to 0$ and there is no combustion. If, however,

$$\mathcal{D} = \mathcal{D}' \exp(\theta/T_*) \quad , \tag{2}$$

where $\mathcal{D}'$ only depends algebraically on θ, nontrivial results can be obtained. (Since $\mathcal{D}$ is to be found, we can make whatever assumptions about it lead to results.) The reaction term is now

$$\Omega = \mathcal{D}'Y \exp(\theta/T_*-\theta/T) \tag{3}$$

and three possibilities arise.

If T is smaller than T_*, then Ω is exponentially small in the limit $\theta \rightarrow \infty$; the reaction is said to be frozen. If T is greater than T_*, then

$$Y = 0 \tag{4}$$

to all (algebraic) orders, since Ω cannot be exponentially large; all the reactant has been consumed, so that equilibrium prevails. If T and T_* differ by $O(\theta^{-1})$, Ω is algebraic in θ; there is active reaction. Thus, we may expect a combustion field containing no maximum temperature to be divided into a frozen region with $T < T_*$ and an equilibrium region with $T > T_*$ by a surface, called a flame sheet, on which $T = T_*$. The problem is to match expansions valid in the reaction zone of thickness $O(\theta^{-1})$ near the surface with those on the two sides.

2. One-Reactant Model

Consider a single reactant decomposing into products, i.e. the reaction

$$R \rightarrow P. \tag{5}$$

Under various assumptions /4, Ch. 1 and p. 22/, none essential, the governing equations for the plane flame of figure 1 are then

$$\mathcal{L}(T) = -\mathcal{L}(Y) = \Omega, \tag{6}$$

where

$$\mathcal{L} \equiv d/dx - d^2/dx^2 \tag{7}$$

is the convection-diffusion operator and Ω has the definition (1). The equations have been nondimensionalized, temperature being measured in units of Q/c_p and distance in units of $\lambda/c_p M$. As a consequence the laminar-flame eigenvalue

$$\mathcal{D} = DM^{-2} \tag{8}$$

contains the burning rate M.

For simplicity we have taken

$$Le = 1; \tag{9}$$

otherwise the operators on T and Y would be different. Their identity results in the reactionless equation

$$\mathcal{L}(H) = 0 \tag{10}$$

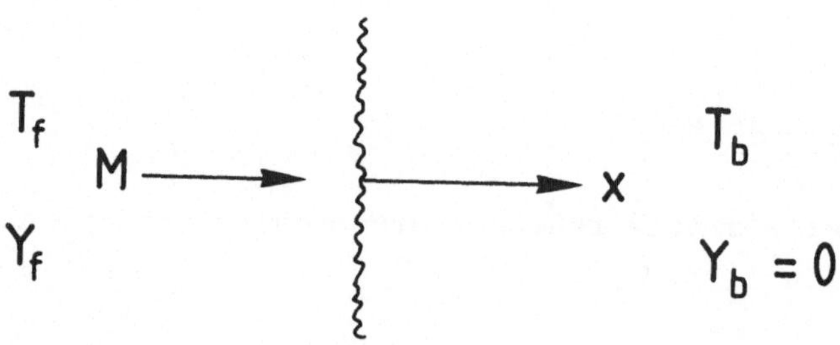

Fig. 1: Notation for plane flame.

for the total enthalpy

$$H = T + Y , \tag{11}$$

which is also called a Shvab-Zeldovich variable. It follows that H is conserved, i.e.

$$T + Y = T_f + Y_f \equiv H_f , \tag{12}$$

a relation that is graphed in figure 2: as T increases from its value T_f in the fresh mixture to its value T_b in the burnt mixture, Y decreases from Y_f to 0. There remains the single equation

$$\mathcal{L}(T) = \mathcal{D} Y \exp(-\theta/T) \tag{13}$$

for the temperature, since Y is now known as a (linear) function of T. The solution of this equation under the boundary conditions in figure 1 is the steady state for problem a (albeit in different units and for Le = 1).

In the limit $\theta \rightarrow \infty$, the temperature on the two sides of the flame sheet becomes

$$T = \begin{cases} T_f + Y_f e^x & \text{for } x < 0, \\ T_b = H_f & \text{for } x > 0 \end{cases} \tag{14}$$

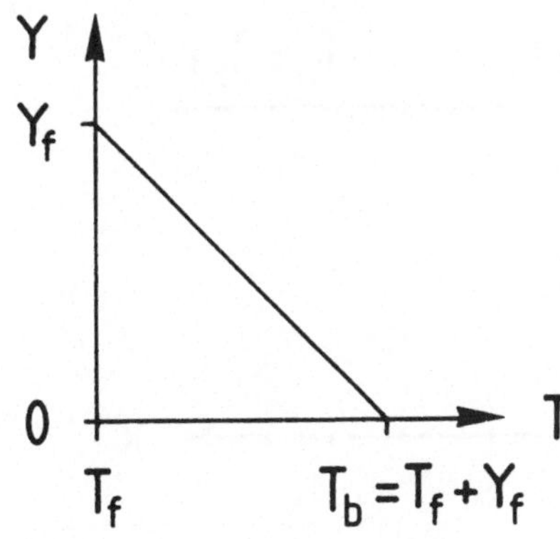

Fig. 2: The T, Y-relation (12).

if x is measured from the sheet. The first of these follows from the solutions 1 and e^x of the frozen-reaction equation $\mathcal{L}(T) = 0$, while the second is obtained on setting Y = 0 in the T, Y-relation (12). (A special argument /4, p.25/ is required to show that equilibrium holds behind the flame sheet even though the temperature does not exceed its value in the reaction zone.) The temperature and mass-fraction profiles are drawn in figure 3.

The initial conditions for problem A can now be recognized as the steady state for $\theta \rightarrow \infty$, at least when Le = 1. They are, therefore, a good starting point for computing the steady-state profiles when θ is large, as has already been mentioned under item (i) in the Introduction.

So far we have not considered the chemical reaction, which must determine the burning rate via $\mathcal{D}$. We do know that it takes place at the $T_b = H_f$, which implies $T_* = T_b$ in the exponential factor of $\mathcal{D}$. Since the thickness of the reaction zone is $O(\theta^{-1})$, the appropriate variable to describe its structure is

$$\xi = \theta x ; \tag{15}$$

then the expansion

$$1/T = 1/T_b + \theta^{-1}\phi(\xi) + \dots \tag{16}$$

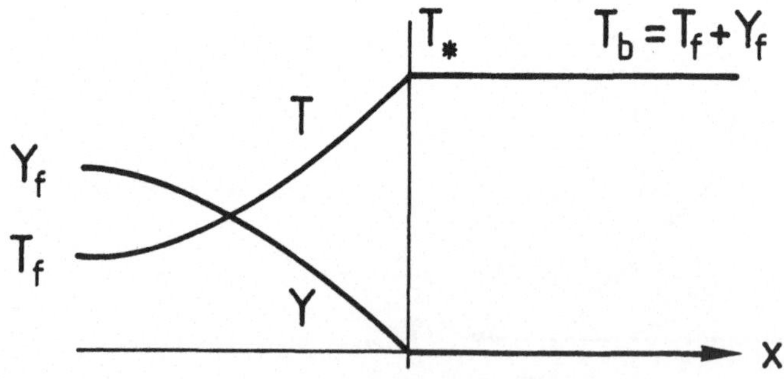

Fig. 3: Asymptotic profiles of T and Y.

in equation (13) leads to

$$d^2\phi/d\xi^2 = \tilde{\mathcal{D}}\phi e^{-\phi} \quad \text{with } \tilde{\mathcal{D}} = \mathcal{D}'\theta^{-2} . \tag{17}$$

The convection term has dropped out, leaving a balance between diffusion and reaction. The algebraic dependence of $\mathcal{D}$ on θ is now determined: $\tilde{\mathcal{D}}$ is independent of θ, so that $\mathcal{D}'$ must be proportional to θ^2. (The same conclusion comes from considering the flame sheet to be a distribution of heat sources; then $\int \Omega dx = O(1)$ implies $\Omega = O(\theta)$ which, together with $Y = O(\theta^{-1})$, gives $\mathcal{D}' = O(\theta^2)$.)

To match with the temperature fields outside the reaction zone, the relevant solution of the structure equation (17) must satisfy the boundary conditions

$$d\phi/d\xi \longrightarrow \begin{cases} -Y_f/T_b^2 & \text{as } \xi \to -\infty, \\ 0 & \text{as } \xi \to +\infty. \end{cases} \tag{18}$$

Only for the particular value

$$\tilde{\mathcal{D}} = Y_f^2/2T_b^2 \tag{19}$$

is it able to do so, and $\tilde{\mathcal{D}}$ is thereby determined. From it we find

$$M = \sqrt{2D} \quad T_b^2 \exp(-\theta/2T_b)/\theta Y_f , \tag{20}$$

a formula obtained by Zeldovich and Frank-Kamenetskii /5/ without formally using activation-energy asymptotics. (When the Lewis number is retained, the factor $\sqrt{2D}$ is replaced by $\sqrt{2DLe}$.) The present derivation is essentially due to Bush and Fendell /6/, who also calculated the next term, of relative order θ^{-1}. Their formula is in remarkably good agreement with computed results for quite moderate values of θ, whatever the value of Le.

The unit of velocity in problem a is M/ρ_f, so that the computed flame velocities should be close to 1 when β is large.

3. Two-Reactant-Model

When an oxidant and a fuel react to form a product, i.e. the reaction is

$$O + F \longrightarrow P \ , \tag{21}$$

the governing equations become

$$\mathcal{L}(T) = -\mathcal{L}(2X) = -\mathcal{L}(2Y) = \Omega \tag{22}$$

where

$$\Omega = \mathcal{D} XY \exp(-\theta/T) \ . \tag{23}$$

For simplicity, we have assumed that both Lewis numbers are 1, so that both Shvab-Zeldovich variables

$$G = T + 2X, \quad H = T + 2Y \tag{24}$$

are constant:

$$T+2X = T_f+2X_f = G_f, \quad T+2Y = T_f+2Y_f = H_f \ . \tag{25}$$

The functions G and H are conditional enthalpies:

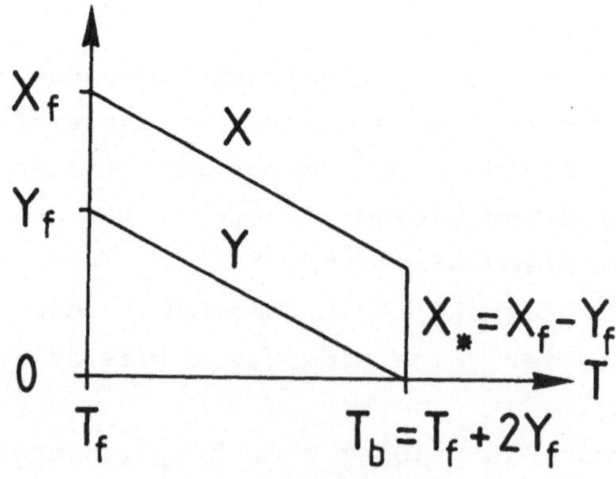

Fig. 4: The T, X, Y-relations (25).

the total enthalpy is G if the oxidant is consumed, i.e. in a fuel-rich mixture; likewise, it is H in a fuel-lean mixture. For definiteness we shall suppose the mixture is fuel-lean, i.e.

$$X_f - Y_f = X_* > 0 . \tag{26}$$

Figure 2 is now replaced by figure 4; when Y has decreased to 0 (in the burnt mixture) the amount X_* of oxidant remains.

Since X and Y are now known as (linear) functions of T, the temperature equation

$$\mathcal{L}(T) = (\mathcal{D}X_*/2)(2Y)\ \exp(-\theta/T) \tag{27}$$

is all that remains to be solved. (In activation-energy asymptotics, the reaction term is only used at the flame sheet, hence the replacement of X by X_*.) The equations (25b) and (27) are identical to the one-reactant equations (12) and (13) under the transformation

$$\mathcal{D}X_*/2,\ 2Y \rightarrow \mathcal{D},\ Y ; \tag{28}$$

we are effectively dealing with a single reactant of double the concentration of the fuel and a modified Damköhler number. The result (20) therefore leads immediately to the burning-rate formula

$$M = \sqrt{DX_*}\ T_b^2\ \exp(-\theta/2T_b)/2\theta Y_f\ . \tag{29}$$

If Y_f is increased, so that the fuel and oxidant are brought closer to stoichiometry, T_b increases and so does the burning rate, except very close to stoichiometry; there the approach of X_* to 0 plays a role in the formula. The analysis breaks down, however, when Y_f is within $O(\theta^{-1})$ of X_f, i.e.

$$X_f - Y_f = \theta^{-1}\ x_* ; \tag{30}$$

in the reaction zone X is now $O(\theta^{-1})$ and cannot be replaced

$X_* = \theta^{-1}x_*$, since that does not properly represent its variations there. Instead of equation (27) we must deal with the original temperature equation where, of course, the T, X, Y-relations (25) are still applicable.

The structure problem now becomes /4, p.36/

$$d^2\phi/d\xi^2 = \tilde{\mathscr{D}} X_1 Y_1 e^{-\phi} \quad \text{with} \; \tilde{\mathscr{D}} = \mathscr{D}'/\theta^3 T_b^2, \quad X_1 - x_* = Y_1 = T_b^2 \phi/2 \qquad (31)$$

and the same boundary conditions (18). The solution leads to the burning-rate formula

$$M = \sqrt{D(x_* + T_b^2)}\; T_b^2 \exp(-\theta/2T_b)/2\theta^{3/2}\; Y_f\;, \qquad (32)$$

which gives an acceptable value at stoichiometry, i.e. for $x_* = 0$. Note the agreement with the result (29) when $x_* = \theta X_* \gg 1$.

Generalizations of the model have been considered by several authors, see for example /7-9/. In particular, Ludford and Sen took Lewis numbers (of fuel and oxidant) different from 1 and each other /10-11/; and extended the theory to non-dilute mixtures /12/, which involves the multicompoment diffusion laws instead of the generalized Fick's law used here.

4. Complex Kinetics

There are already at least three ways in which more detailed kinetics have been considered. In addition, current investigations of complex kinetics give every indication of forming one of the major thrusts in the application of activation-energy asymptotics. Only the governing equations for these three examples will be written down; to explain the analysis and results would add little to our presentation of basic ideas.

Dissociation of the product in the two-reactant model has been taken into account by Ludford and Sen, not only for a dilute mixture /13, 14/ but also for a non-dilute one /15/, by replacing the reaction term (23) with

$$\Omega = \mathcal{D}(XY-Z)\ \exp(-\theta/T)\ . \tag{33}$$

Their main conclusion is that dissociation has only a subsidiary effect on the burning rate, whose characteristics are established by mass diffusion.

Sequential reactions

$$A \rightarrow B \rightarrow C \tag{34}$$

governed by the equations

$$\mathcal{L}(T) = Q_1\Omega_1+Q_2\Omega_2,\ \mathcal{L}(X) = -\Omega_1,\ \mathcal{L}(Y) = \Omega_1-\Omega_2 \tag{35}$$

with

$$\Omega_1 = \mathcal{D}_1 X\ \exp(-\theta_1/T),\ \Omega_2 = \mathcal{D}_2 Y\ \exp(-\theta_2/T) \tag{36}$$

have been considered by Kapila and Ludford /16/. Now there are two large parameters, θ_1 and θ_2, and their relative size plays a role. The problem has recently been re-examined by Margolis and Matkowsky /17/.

Zeldovich /18/ has introduced an autocatalytic chain reaction

$$A+B \rightarrow 2B\ ,\qquad B+B \rightarrow C \tag{37}$$

governed by the equations

$$\mathcal{L}(T) = \Omega_2,\ \mathcal{L}(X) = -\Omega_1,\ \mathcal{L}(Y) = \Omega_1-\Omega_2 \tag{38}$$

with

$$\Omega_1 = \mathcal{D}_1 XY\ \exp(-\theta/T),\quad \Omega_2 = \mathcal{D}_2 Y^2\ . \tag{39}$$

Only the first step (radical production) has non-zero activation energy and only the second step (chain-breaking recombination) releases heat. Linán /19/ has shown that, for large values of θ, there are three essentially different types of structure, depending on the relative magnitudes of $\mathcal{D}_1$ and $\mathcal{D}_2$. (This problem shows that not all the steps in a reaction need have large activation energies for the method to apply; simplification rather than solution is obtained by activation-energy asymptotics.)

While it can hardly be claimed that these investigations of more complex kinetics tell us anything that was not already known about problems such as b, they are not far away from doing so. /See item (iii) in the Introduction./ The main advantage that analytical methods have over numerical ones is that results are obtained for all parameters values, so that the effects of various processes can be separated. That is where activation-energy asymptotics can be expected to make a contribution.

Nomenclature
(not defined by context)

c_p	specific heat at constant pressure
D	Damköhler number (proportional to rate constant)
Le	Lewis number
Q	heat of combustion
M	mass flux through reaction zone
T	dimensionless temperature
X,Y	mass fractions of reactants
Z	mass fraction of product
β	dimensionless activation energy in problem a
θ	dimensionless activation energy used here
λ	coefficient of thermal conductivity
ρ	density

References

/1/ F.A. Williams: Asymptotic methods in deflagrations. In First International Specialists Meeting of the Combustion Institute, pp. LXX-LXXIV. Section Française du Combustion Institute, 1981.

/2/ F.A. Williams: Combustion Theory. Reading, Massachusetts: Addison-Wesley, 1965.

/3/ N. Peters: Theoretical implications of nonequal diffusivities of heat and matter on the stability of a plane premixed flame. This volume.

/4/ J.D. Buckmaster & G.S.S. Ludford: Theory of Laminar Flames. Cambridge: University Press, 1981.

/5/ Y.B. Zeldovich & D.A. Frank-Kamenetskii: A theory of thermal propagation of flame. Acta Physicochimica U.R.S.S., **9**, 341-50 (1938).

/6/ W.B. Bush & F.E. Fendell: Asymptotic analysis of laminar flame propagation for general Lewis numbers. Combustion Science and Technology, **1**, 421-8 (1970).

/7/ J.F. Clarke: The pre-mixed flame with large activation energy and variable mixture strenth: elementary asymptotic analysis. Combustion Science and Technology, **10**, 189-94 (1975).

/8/ G.F. Carrier, F.E. Fendell & W.B. Bush: Stoichiometry and flame-holder effects on a one-dimensional flame. Combustion Science and Technology, **18**, 33-46 (1978).

/9/ T. Mitani: Propagation velocities of two-reactant flames. Combustion Science and Technology, **21**, 175-7 (1980).

/10/ A.K. Sen & G.S.S. Ludford: The near-stoichiometric behavior of combustible mixtures. Part I: Diffusion of the reactants. Combustion Science and Technology, **21**, 15-23 (1979).

/11/ G.S.S. Ludford & A.K. Sen: Burning rate maximum of a plane premixed flame. Progress in Astronautics and Aeronautics, **76**, 427-36 (1981). (Combustion in Reactive Systems, ed. J. Ray Bowen, N. Manson, Antoni K. Oppenheim, and R.I. Soloukhin.)

/12/ A.K. Sen & G.S.S. Ludford: Effects of mass diffusion on the burning rate of non-dilute mixtures. In Eighteenth International Symposium on Combustion, pp. 417-24. Pittsburgh: The Combustion Institute, 1981.

/13/ A.K. Sen & G.S.S. Ludford: The near-stoichiometric behavior of combustible mixtures. Part II: Dissociation of the products. Combustion Science and Technology, **26**, 183-91 (1981).

/14/ A.K. Sen & G.S.S. Ludford: Maximum flame temperature and burning rate of combustible mixtures. Submitted for publication.

/15/ G.S.S. Ludford & A.K. Sen: The effect of dissociation on the near-stoichiometric burning of non-dilute mixtures. Submitted for publication.

/16/ A.K. Kapila & G.S.S. Ludford: Two-step sequential reactions for large activation energy. Combustion and Flame, **29**, 167-76 (1977).

/17/ S.B. Margolis & B.J. Matkowsky: Flame propagation with a sequential reaction mechanism. SIAM Journal on Applied Mathematics (to appear).

/18/ Y.B. Zeldovich: Theory of flame propagation. NACA Technical Memorandum 1282, 1951. Translation of: K Teorii Rasprostranenia Plameni. Zhurnal Fizicheskoi Khimii, **22**, 27-49 (1948).

/19/ A. Liñán: A theoretical analysis of premixed flame propagation with an isothermal chain reaction. Instituto Nacional de Tecnica Aerospacial "Esteban Terradas", USAF Contract No. E00AR 68-0031, Technical Report No. 1, 1971.

THEORETICAL IMPLICATIONS OF NONEQUAL DIFFUSIVITIES OF HEAT AND MATTER ON THE STABILITY OF A PLANE PREMIXED FLAME

by

N. Peters

Institut für Allgemeine Mechanik
Rheinisch-Westfälische Technische Hochschule Aachen
Aachen, West Germany

ABSTRACT

Sivashinsky's large activation energy analysis of the stability characteristics of the plane premixed flame with Lewis number different from unity is adapted to test problem A of the workshop on "Numerical Methods in Laminar Flame Propagation".

1. INTRODUCTION

The structure of a premixed flame with one-step irreversible kinetics may be divided into three parts: A preheat zone, a reaction zone and an equilibrium zone.

If the reaction has a large activation energy β, the reaction zone is thin of order $1/\beta$ compared to the preheat zone. If the diffusivities of heat and matter are equal, i.e. the Lewis number is equal to one, the enthalpy within the flame is constant. If, however, the diffusivities of heat and the reactant are different, a non-constant enthalpy profile will result. For instance, if the diffusivity of heat exceeds that of the reactant, $Le > 1$, heat will diffuse at a faster rate from the reaction zone into the preheat zone than the reactant - and its enthalpy of formation - diffuses into reaction zone. Therefore the enthalpy in the reaction zone will decrease. This leads to a decrease of temperature in the reaction zone and, since the flame velocity depends strongly on temperature, to a decrease in flame velocity. The same argument leads to an increase in flame velocity for $Le < 1$.

Sivashinsky /1/ has shown by an asymptotic analysis for large activation energy that the unsteady flame with $Le > 1$ is unstable while the one with $Le < 1$ is stable. An underlying assumption of his treatment is that the flame is slowly varying, i.e. that the time derivatives in the equations for the preheat zone are of order $1/\beta$ compared to the convective and diffusive terms. This analysis will be repeated for the case of test problem A. The deficiencies due to the assumption of a very large activation energy will be discussed.

2. FORMULATION

The test problem A is defined by the following set of equations

$$\frac{\partial T}{\partial t} = \frac{\partial^2 T}{\partial x^2} + R \tag{1}$$

$$\frac{\partial Y}{\partial t} = \frac{1}{Le}\frac{\partial^2 Y}{\partial x^2} - R \tag{2}$$

$$R = \frac{\beta^2}{2Le} Y \exp\left(-\frac{\beta(1-T)}{1-\alpha(1-T)}\right) \quad . \tag{3}$$

We shall introduce a coordinate system that is attached to the flame surface

$$\xi = x - x_F(t) \tag{4}$$

where $x_F(t)$ is the position of the thin reaction zone. The assumption of a slowly moving flame leads to the new time coordinate

$$\tau = \varepsilon t \tag{5}$$

where ε is a small parameter that will be determined in the inner zone analysis (Chapter 4). The new time coordinate τ is of order unity. The transformation rules are

$$\frac{\partial}{\partial x} = \frac{\partial}{\partial \xi}$$

$$\frac{\partial}{\partial t} = v_F \frac{\partial}{\partial \xi} + \varepsilon \frac{\partial}{\partial \tau} \tag{6}$$

where

$$v_F \equiv - \frac{\partial x_F}{\partial t} \tag{7}$$

is a function of τ. Introducing this into eqs. (1) and (2) leads to

$$\varepsilon \frac{\partial T}{\partial \tau} + v_F \frac{\partial T}{\partial \xi} = \frac{\partial^2 T}{\partial \xi^2} + R \tag{8}$$

$$\varepsilon \frac{\partial Y}{\partial \tau} + v_F \frac{\partial Y}{\partial \xi} = \frac{1}{Le} \frac{\partial^2 Y}{\partial \xi^2} - R \quad . \tag{9}$$

One can combine eqs. (8) and (9) to obtain an equation for the enthalpy $H \equiv T + Y - 1$

$$\varepsilon \frac{\partial H}{\partial \tau} + v_F \frac{\partial H}{\partial \xi} = \frac{\partial^2 H}{\partial \xi^2} + (\frac{1}{Le} - 1) \frac{\partial^2 Y}{\partial \xi^2} \quad . \tag{10}$$

3. THE PREHEAT ZONE STRUCTURE

In the limit of large activation energy the reaction takes place only within the very thin reaction zone. Therefore the preheat zone may be analysed by setting $R = 0$. The effect of nonequal diffusivities leads to a perturbation of the preheat zone structure. We therefore introduce the expansion

$$\begin{aligned} T &= T^o + \varepsilon T^1 + O(\varepsilon^2) \\ Y &= Y^o + \varepsilon Y^1 + O(\varepsilon^2) \\ H &= H^o + \varepsilon H^1 + O(\varepsilon^2) \\ v_F &= v_F^o + \varepsilon v_F^1 + O(\varepsilon^2) \quad . \end{aligned} \tag{11}$$

Introducing this into eqs. (8) and (9) yields for $\varepsilon \to 0$ to zeroth order

$$v_F^o \frac{\partial T^o}{\partial \xi} = \frac{\partial^2 T^o}{\partial \xi^2} \tag{12}$$

$$v_F^o \frac{\partial Y^o}{\partial \xi} = \frac{1}{Le} \frac{\partial^2 Y^o}{\partial \xi^2} \quad . \tag{13}$$

Introducing the boundary conditions of test problem A

$$\xi \to -\infty : \qquad T^o = 0 \, , \qquad Y^o = 1 \tag{14}$$

and the conditions at the flame surface

$$\xi = 0 \quad : \qquad T^o = 1 \, , \qquad Y^o = 0 \tag{15}$$

leads to the solution of the zeroth order temperature and concentration

$$T^o = \exp(v_F^o \xi) \tag{16}$$

$$Y^o = 1 - (T^o)^{Le} \quad . \tag{17}$$

The zeroth order enthalpy is therefore given by

$$H^o = \exp(v_F^o \xi) - \exp(v_F^o Le\, \xi) \quad . \tag{18}$$

The first order enthalpy equation is

$$\frac{\partial H^o}{\partial \tau} + v_F^o \frac{\partial H^1}{\partial \xi} + v_F^1 \frac{\partial H^o}{\partial \xi} = \frac{\partial^2 H}{\partial \xi^2} + \left(\frac{1}{Le} - 1\right) \frac{\partial^2 Y^1}{\partial \xi^2} \quad . \tag{19}$$

Integrating this equation between $\xi = -\infty$ and $\xi = 0_+$ yields

$$\frac{\partial}{\partial \tau} \int_{-\infty}^{0} H^o d\xi + v_F^o(H^1(0_+) - H^1(-\infty)) + v_F^1(H^o(0_+) - H^o(-\infty))$$
$$= \frac{\partial H}{\partial \xi}\Big|_{0_+} - \frac{\partial H}{\partial \xi}\Big|_{-\infty} + \left(\frac{1}{Le} - 1\right)\left(\frac{\partial Y^1}{\partial \xi}\Big|_{0_+} - \frac{\partial Y^1}{\partial \xi}\Big|_{-\infty}\right) \quad . \tag{20}$$

Since the gradients of enthalpy and concentration vanish at 0_+ and at $-\infty$, the r.h.s. of eq. (20) is zero. Using eq. (18), one obtains

$$\left(1 - \frac{1}{Le}\right) \frac{d}{d\tau} \left(\frac{1}{v_F^o}\right) + v_F^o(H^1(0_+) - H^1(-\infty)) = 0 \quad . \tag{21}$$

This equation represents a balance between the accumulation of enthalpy in the preheat zone and the difference between inflow and outflow of enthalpy. Since the enthalpy inflow $H^1(-\infty) = 0$, nonequal diffusivities lead to a non-zero enthalpy outflow at $\xi = 0$. Here the first order enthalpy is equal to the first order temperature

$$H^1(0) \equiv T^1(0) + Y^1(0) = T^1(0) \tag{22}$$

since the concentration at $\xi = 0$ is zero to all orders. Finally, one obtains

$$(1 - \frac{1}{Le}) \frac{dv_F^o}{d\tau} = v_F^{o^3} T^1(0) \quad . \tag{23}$$

This is a first relation between v_F^o and $T^1(0)$. By deriving a second relation from the inner zone analysis, we will be able to eliminate $T^1(0)$ and obtain a differential equation for v_F^o.

4. INNER ZONE ANALYSIS

The reaction zone has a time-independent structure that must be matched to the outer, slowly time-dependent structure. As it is thin, a stretched coordinate is introduced

$$\zeta = \xi/\varepsilon \quad . \tag{24}$$

The dependent variables are expanded around their values at $\xi = 0$

$$\begin{aligned} T &= 1 + \varepsilon\Theta(\zeta) \\ Y &= \varepsilon Le y(\zeta) \end{aligned} \tag{25}$$

Introducing these into eqs. (8) and (9) leads to first order

$$\frac{d^2\Theta}{d\zeta^2} = \varepsilon R(\Theta,y) \tag{26}$$

$$\frac{d^2Y}{d\zeta^2} = \varepsilon R(\Theta,y) \tag{27}$$

when the convective term is of lower order. Eqs. (26) and (27) can be combined

$$\frac{d^2\Theta}{d\zeta^2} = \frac{d^2Y}{d\zeta^2} \quad . \tag{28}$$

The boundary conditions obtained from matching with the outer solution are

$$\zeta \to \infty : \qquad \Theta = T^1(0) \ , \qquad Y = 0 \ , \qquad \frac{d\Theta}{d\zeta} = \frac{dy}{d\zeta} = 0 \tag{29}$$

Integration of eq. (28) leads therefore to

$$\Theta + y = T^1(0) \quad . \tag{30}$$

The reaction rate R is written in terms of Θ and y as

$$R = \frac{\beta^2 \varepsilon y}{2} \exp \frac{\varepsilon\beta\Theta}{1+\varepsilon\Theta} \quad . \tag{31}$$

The term in the exponent yields a suitable definition for the yet unknown small quantity ε

$$\varepsilon = \beta^{-1} \quad . \tag{32}$$

Then, in the limit $\varepsilon \to 0$, $\beta \to \infty$ one obtains with eqs. (27) and (30) for the concentration equation in the reaction zone

$$\frac{d^2y}{d\zeta^2} = \frac{1}{2}\, y \exp T^1(0) \exp(-y) \quad . \tag{33}$$

This equation can be written as

$$\frac{dp^2}{dy} = y \exp T^1(0) \exp -y \quad . \tag{34}$$

where $p = dy/d\zeta$. Integrating once between ζ and $\zeta = +\infty$ or y and $y = 0$, respectively, yields with the boundary conditions (29)

$$p^2 = \exp T^1(0) \int_0^y y' \exp -y' dy' \quad . \tag{35}$$

At $\zeta = -\infty$, $y = \infty$ one obtains for the gradient

$$\left.\frac{dy}{d\zeta}\right|_{-\infty} = \exp \frac{1}{2} T^1(0) \quad . \tag{36}$$

The matching condition between the preheat zone and the reaction zone is

$$\lim_{\xi \to 0_-} \frac{\partial Y^o}{\partial \xi} = \text{Le} \lim_{\zeta \to -\infty} \frac{dy}{d\zeta} \quad . \tag{37}$$

With eqs. (16) and (17) one obtains

$$\left.\frac{\partial Y^o}{\partial \xi}\right|_{0_-} = \text{Le}\ v_F^o \quad . \tag{38}$$

Therefore a second relation between the flame velocity v_F^o and $T^1(0)$ is obtained as

$$v_F^o = \exp\left(\frac{1}{2}\,T^1(0)\right) \quad . \tag{39}$$

5. DISCUSSION

Combining eqs. (23) and (39) one obtains the differential equation

$$\frac{dv_F^o}{d\tau} = \frac{\text{Le}}{(\text{Le} - 1)}\, 2\, {v_F^o}^3 \ln v_F^o \quad . \tag{40}$$

There are two steady state solutions, namely $v_{F,1}^o = 0$ and $v_{F,2}^o = 1$. Depending on the value of the Lewis number only one of these steady state solutions is stable. For instance, if $\text{Le} < 1$ an initial value $0 < v_F^o < 1$ will increase until $v_F^o = 1$ while an initial value of $v_F^o > 1$ will decrease until $v_F^o = 1$. Therefore the steady state solution $v_{F,2}^o$ is stable for $\text{Le} < 1$. On the other hand for $\text{Le} > 1$ and an initial value $0 < v_F^o < 1$ the flame velocity will decrease to $v_F^o = 0$. This decrease indicates that the flame will extinguish. For an initial value $v_F^o > 1$ the flame velocity will increase to infinity. Therefore in the case $\text{Le} > 1$ the steady state solution $v_{F,2}^o = 1$ is unstable to small perturbations.

However, the numerical solutions for $Le = 2$, $\beta = 20$ of test problem A show oscillations of the flame velocity with very high peaks. After an initial decrease in flame velocity there is a sharp increase again. This seems to correspond to a situation where enthalpy has been accumulated in the preheat zone during the period of decreasing flame velocity. Only little heat from the flame zone is necessary to ignite this mixture. The reason that the analysis above does not predict this behaviour is due to the fact that the limit of an infinitely large activation energy has been employed. A stability analysis of the problem based on a finite activation energy should - if it could be done correctly - predict the oscillations observed in the numerical calculations.

REFERENCES

/1/ Sivashinsky, G.I.: On a Converging Spherical Flame Front, Int. J. Heat Mass Transf. 17 (1974), 1499-1505.

THE EFFECT OF LEWIS NUMBER GREATER THAN UNITY ON AN UNSTEADY PROPAGATING FLAME WITH ONE - STEP CHEMISTRY

by

B. Rogg
Institut für Allgemeine Mechanik
Rheinisch-Westfälische Technische Hochschule Aachen
Aachen, West Germany

ABSTRACT

The effect of unequal diffusivities of heat and reactant in a one-dimensional unsteady propagating flame with one-step chemistry is studied numerically. It is shown that the governing equations may have steady oscillating solutions, depending upon the values of Lewis number and nondimensional activation energy. Furthermore, a numerical method for the solution of the basic equations is indicated.

1. INTRODUCTION

A very convenient assumption, often introduced in analytical flame studies, is that of equal diffusivities of heat and reactants, $Le_i = \lambda/\rho c_p D_i \equiv 1$. However, as was already pointed out by Lewis and von Elbe /1/, /2/ for a Lewis number larger than one the rate of heat conduction from the burnt gases into the preheat zone exceeds the rate of diffusion at which the reactants are transported out of it leading to an excess enthalpy.

However, they did not recognize that this effect will also lead to an enthalpy decrease in the thin reaction zone close to the adiabatic flame temperature. This implies a decrease in temperature and due to its strong sensitivity to the temperature the flame velocity is decreased. Sivashinsky /3/ (cf. also Peters' adaption of his result to the test problem /4/) showed that in the limit of a large activation energy and $Le > 1$ the flame velocity decreases monotonically to zero.

In the present paper the case $Le > 1$ is studied numerically for an unsteady propagating flame with one-step chemistry. There are two reasons for doing this: i) Up to now no systematical investigation exists for the case $Le > 1$ and finite values of the activation energy; and ii) If such an investigation brings to light difficulties for the numerical solution of the corresponding differential equations the question for an appropriate numerical method will arise. For the second reason the equations of a test problem proposed for the GAMM-Workshop at Technical University Aachen, Germany, Oct. 12-14, 1981, /5/ are solved. The flame velocity is evaluated as a function of time, varying the values of Lewis number Le and nondimensional activation energy β. As a result a limit line separating the steady from the oscillating regime for the flame velocity is obtained and an attempt is made to explain the occurrence of observed flame velocity oscillations.

2. BASIC EQUATIONS

The nondimensional governing equations for an unsteady propagating flame with one-step chemistry and Lewis number different from unity /5/ are:

$$\frac{\partial T}{\partial t} = \frac{\partial^2 T}{\partial x^2} + R \tag{1}$$

$$\frac{\partial Y}{\partial t} = \frac{1}{Le}\frac{\partial^2 Y}{\partial x^2} - R \quad . \tag{2}$$

T and Y denote normalized temperature and normalized mass fraction of the reactant, respectively. The nondimensional reaction rate is given by

$$R = \frac{\beta^2}{2Le} Y \exp\left(- \frac{\beta(1-T)}{1-\alpha(1-T)}\right) \quad . \tag{3}$$

β denotes the nondimensional activation energy. The initial and boundary conditions are:

initial conditions: $t = 0$

$$\begin{aligned} x \leq 0 : \quad & T = \exp x \\ & Y = 1 - \exp(Le\ x) \\ x \geq 0 : \quad & T = 1 \\ & Y = 0 \end{aligned} \tag{4}$$

boundary conditions: $t > 0$

$$\begin{aligned} x \to -\infty : \quad & T = 0 \\ & Y = 1 \\ x \to +\infty : \quad & \frac{dT}{dx} = 0 \\ & \frac{dY}{dx} = 0 \quad . \end{aligned} \tag{5}$$

3. NUMERICAL SCHEME

From a numerical point of view it is appropriate to consider the problem in a new (ξ,τ)-coordinate system, which is moving with the flame front. Thus, the coordinate transformation

$$\xi = x - x_f(t)$$
$$\tau = t \tag{6}$$

is introduced into eqs. (1) and (2). Here x_f denotes the location of the flame front and the flame velocity v_f is defined as

$$v_f = - \frac{dx_f}{dt} \quad . \tag{7}$$

Writing again x and t instead of ξ and τ, eqs. (1) and (2) result in

$$\frac{\partial T}{\partial t} + v_f \frac{\partial T}{\partial x} = \frac{\partial^2 T}{\partial x^2} + R \tag{8}$$

$$\frac{\partial Y}{\partial t} + v_f \frac{\partial Y}{\partial x} = \frac{1}{Le} \frac{\partial^2 Y}{\partial x^2} - R \quad , \tag{9}$$

while the initial conditions (4) and boundary conditions (5) remain unchanged. To solve eqs. (8) and (9) the method of lines /6/ turns out to be the most effective. Thus, the interphase software PDEONE /7/ is used to perform the spatial discretization. The resulting system of ordinary differential equations with the time coordinate being the independent variable is integrated using the EPISODE-package /8/. Prior to every new time step an iteration process is started to find that value for the flame velocity which is fixing the flame front to the origin of the coordinate system. The location of the flame front $x_f(t)$ is defined by the arithmetic mean of the temperatures at hot and cold boundary of the flame:

$$T(t, x=x_f(t)) \equiv 0.5 \quad . \tag{10}$$

From definition (10) and the initial temperature profile (4) the initial condition for $x_f(t)$ can be determined.

4. RESULTS AND DISCUSSION

In Fig. 1 the limit between the steady and the oscillating regime for the flame velocity v_f is represented. On the right of the minimum Lewis number $Le = 1.402$ oscillations leading to a limit cycle for time $t \to \infty$ are observed for an increasing range of nondimensional activation energies β. For values of β above and below the limiting curve initial disturbances are damped leading to a steady flame propagation for $t \to \infty$.

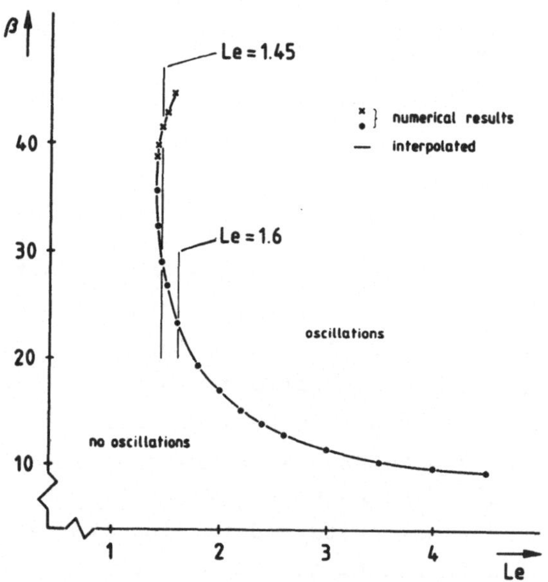

Figure 1: Curve of oscillation limits in the (Le, β)-plane.

In Figs. 2 to 4 the flame velocity v_f is plotted vs. time coordinate t for $Le = 1.45$ and three distinctive values of β (β = 25, 32, 42), each value corresponding to one of the indicated regions. In all cases first the flame velocity decreases from an initial value near one to a minimum value and then increases. This initial decrease occurs because - the Lewis number being greater than unity - thermal diffusivity exceeds molecular diffusivity. Therefore, the rate of heat losses from the reaction zone exceeds the rate at which the reactant is transported into the reaction zone. Hence the enthalpy and consequently the temperature decreases in the reaction zone while both increase in the preheat zone. Since flame velocity depends exponentially on temperature this causes the decrease in flame velocity. Due to the transient nature of the problem the gradients of temperature and reactant concentration flatten as the flame is slowing down. The increase of enthalpy and temperature in the preheat zone results in an accelerated ignition of the reactant at the interphase between preheat zone and reaction zone. This is a velocity increasing effect which counteracts the first. Therefore the flame velocity reaches a minimum and increases again.

If the nondimensional activation energy lies far below the lower oscillation limit, that velocity increasing effect is rather weak and leads only to a slow stabilisation of the temperature and concentration profiles. Thus, in this case the flame velocity increases gradually to its steady state value.

Figure 2 shows the situation for a value of β which lies close below the lower oscillation limit. Here the flame velocity overshoots its steady state value. The oscillations are damped in time because in spite of increasing flame velocity the heat loss due to conduction always compensates the heat production due to reaction. This results in a stabilisation effect on the temperature profile and hence on the concentration profile and the flame velocity.

Figure 3 shows steady oscillations of the flame velocity. In this case the rate of heat loss can not compensate the rate of heat production and the temperature in the reaction zone increases exponentially, causing the reaction rate to do likewise, and a kind of thermal explosions results. Thus, the flame velocity increases sharply until the reactant is consumed in a large

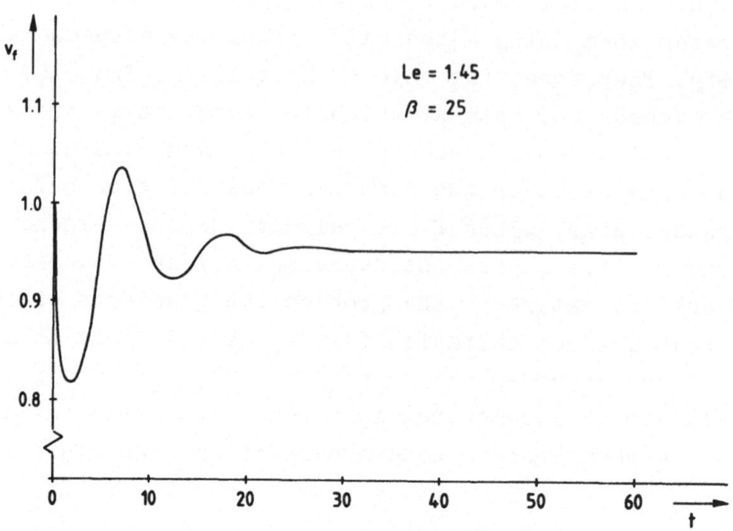

Figure 2: Damped oscillations of flame velocity (Le = 1.45, β = 25) .

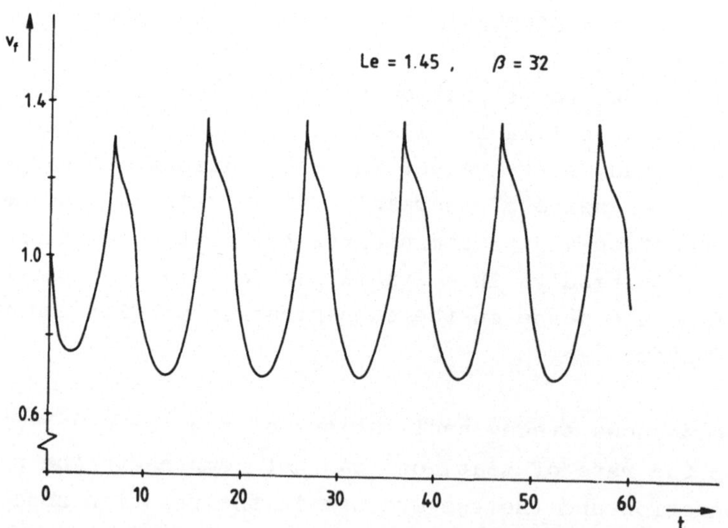

Figure 3: Steady oscillations of flame velocity (Le = 1.45, β = 32) .

region ahead of the flame. Then, heat production rapidly slows down and the temperature in the reaction zone lowers due to heat losses resulting in a rapidly decreasing flame velocity. With time more reactant becomes available and so the reaction is gradually accelerated, although the flame velocity is still decreasing. This cycle recurs continuously.

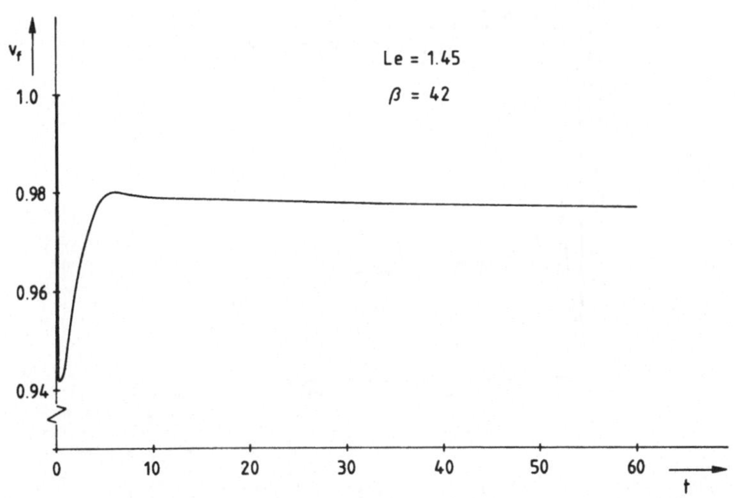

Figure 4: Flame velocity vs. time for values of Le and β above the upper oscillation limit (Le = 1.45, β = 42) .

Figure 4 shows a case in which β lies above the upper oscillation limit. At the moment it is not clear, if a steady state will be reached to $t \to \infty$. It was observed that the required computer time was about three times larger than for cases corresponding to a point below the lower oscillation limit.

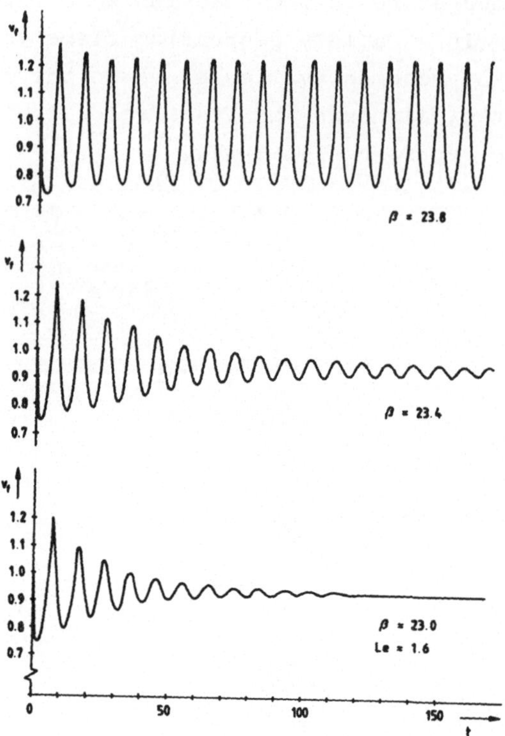

Figure 5: Transition from damped to steady oscillations of flame velocity (Le = 1.6) .

Figure 5 shows the transition from damped to steady oscillations for Le = 1.6 when β is enlarged from 23.0 to 23.8 across the lower oscillation limit.

5. CONCLUSIONS

The influence of Lewis number and activation energy on a one-dimensional unsteady propagating flame with one-step chemistry has been investigated. For the case $Le > 1$ the nondimensional activation energy β was varied in the range from 8 to 50 and numerical evaluations of the normalized flame velocity were performed. A limit line for the flame velocity separating the steady from the oscillating regime in the (Le,β)-plane was calculated. However, it is not yet possible to give a numerical verification of the conclusion /3/, /4/ that in the limit of a large activation energy and $Le > 1$ the flame velocity decreases monotonically to zero.

NOMENCLATURE

c_p	specific heat at constant pressure
D_i	diffusion coefficient of species i
Le_i	Lewis number, ratio of thermal diffusivity to mass diffusivity of species i
R	normalized reaction rate
t	normalized time coordinate
T	normalized temperature
v_f	normalized flame velocity
x	normalized space coordinate
x_f	normalized location of the flame front
Y	normalized mass fraction of reactant

GREEK SYMBOLS

α	parameter used in eq. (3) numerical value: $\alpha = 0.8$

β	nondimensional activation energy
ξ	normalized space coordinate
λ	thermal conductivity
ρ	density
τ	normalized time coordinate

REFERENCES

/1/ Von Elbe, G. and Lewis, B. (1959): Excess Enthalpy and the Initiation and Stability of Combustion Waves. 7th Symp. (Int.) on Comb., The Combustion Institute, London, pp. 342-346.

/2/ Lewis, B. and von Elbe, G. (1961): Combustion, Flames and Explosions of Gases. Academic Press, New York.

/3/ Sivashinsky, G.I. (1974): On a Converging Spherical Flame Front. Int. J. Heat Mass Transfer, Vol. 17, pp.1499-1506.

/4/ Peters, N. (1981): Theoretical Implications of Nonequal Diffusivities of Heat and Matter on the Stability of a Plane Premixed Flame. See this volume.

/5/ GAMM-Workshop on Numerical Methods in Laminar Flame Propagation, Oct. 12-14, 1981, Technical University Aachen, Germany.

/6/ Kurtz, L.A., Smith, R.E., Parks, L.E., Boney, L.R. (1978): A Comparison of the Method of Lines to Finite Difference Techniques in Solving Time-Dependent Partial Differential Equations. Computers and Fluids, Vol. 6, pp. 49-70.

/7/ Sincovec, R.F., Madsen, N.K. (1975): Software for Nonlinear Partial Differential Equations. ACM Transactions on Mathematical Software, Vol. 1, No. 3, pp. 232-263.

/8/ Hindmarsh, A.C., Byrne, G.D. (1977): Episode: An Effective Package for the Integration of Systems of Ordinary Differential Equations. California University, Livermore (USA), Livermore Lab..

DISCUSSION OF TEST PROBLEM B

by

J. Warnatz

Institut für Physikalische Chemie
Technische Hochschule Darmstadt, Darmstadt, W. Germany

Abstract

The results of the eight groups that have calculated test problem B of the GAMM-Workshop on "Numerical Methods in Laminar Flame Propagation" are summarized. The problem is formulated and some background given. In general, there is good agreement of the results calculated by different methods. Minor deviations from the results expected can be attributed either to the transport model or to deficiencies of the grid point system used.

1. Introduction

Test problem B treats the stationary propagation of a one-dimensional (flat) flame front, including complete chemistry by use of a set of elementary reactions, and including realistic transport properties by use of an appropriate multicomponent transport model.

Literature work on the simulation of laminar flame propagation is done by different workers using different reaction mechanisms, various transport models, and a number of numerical methods. This leads to the question how far the results of these calculations are dependent on the choice of the large number of physical and numerical parameters.

Especially, conclusions with regard to the relevance of a reaction mechanism are only possible in the case of disappearing influence of the numerical scheme and of the (necessary) simplifications of the transport model. To determine these influences, calculations by different groups have to be performed using a common specified reaction mechanism.

The special example considered here is a H_2-air freely propagating laminar flat flame. This system is simple enough to allow fast calculations making possible the test of different mechanisms, transport models etc. On the other hand, this flame is complicated enough to be representative for more complicated systems relevant to engineering (e.g. hydrocarbon combustion).

The H_2-O_2-N_2 system nowadays is well known, in particular by the pioneering work of Dixon-Lewis and coworkers (see [1]). There is no problem in calculating concentration-, pressure-, and temperature-dependence of the flame velocity [2]; an example (including the test problem) is given in Fig. 1, comparing measured and calculated flame velocities in H_2-air flames at atmospheric pressure for different unburnt gas mixtures.

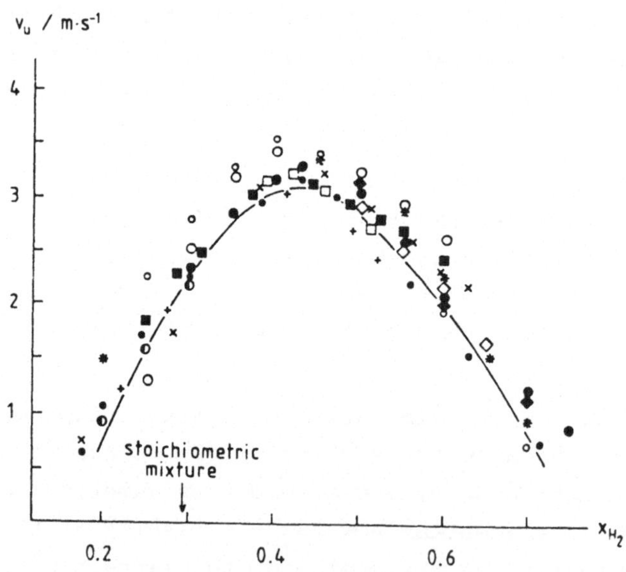

Fig. 1: Measured and calculated flame velocities in H_2-air mixtures ; T_u = 298 K, P = 1 bar. Points: measurements (for reference see [3]); line: calculated with the mechanism given in Table 1.

(The problems connected with the experimental determination of flame velocities in the H_2-O_2-N_2 system are discussed in /5/)

An actual reaction mechanism in the H_2-O_2-N_2 system (neglecting reactions forming and consuming the by-product H_2O_2) is listed in Table 1 [4]; the corresponding flame velocity is 219 cm/s for stoichiometric H_2-air flames at atmospheric pressure (test problem). A similar mechanism is given by Dixon-Lewis resulting in flame velocities of 202 cm/s [5] and 215 cm/s [6].

No.	Reaction	$\frac{A}{(cm^3/mol)^{n-1}s^{-1}}$	b	$\frac{E}{kJ/mol}$
(1)	$H + O_2 \rightarrow OH + O$	$2.2 \cdot 10^{14}$	0	70.3
(2)	$O + OH \rightarrow H + O_2$	$1.8 \cdot 10^{13}$	0	0
(3)	$O + H_2 \rightarrow OH + H$	$1.5 \cdot 10^{7}$	2.00	31.6
(4)	$H + OH \rightarrow O + H_2$	$6.7 \cdot 10^{6}$	2.00	23.3
(5)	$OH + H_2 \rightarrow H_2O + H$	$1.0 \cdot 10^{8}$	1.60	13.8
(6)	$H + H_2O \rightarrow OH + H_2$	$4.6 \cdot 10^{8}$	1.60	77.6
(7)	$OH + OH \rightarrow H_2O + O$	$1.5 \cdot 10^{9}$	1.14	0
(8)	$H_2O + O \rightarrow OH + OH$	$1.5 \cdot 10^{10}$	1.14	72.1
(9)	$H + H + M \rightarrow H_2 + M$	$9.7 \cdot 10^{16}$	-0.60	0
(10)	$H + OH + M \rightarrow H_2O + M$	$2.2 \cdot 10^{22}$	-2.00	0
(11)	$H + O_2 + M \rightarrow HO_2 + M$	$2.0 \cdot 10^{18}$	-0.80	0
(12)	$HO_2 + M \rightarrow H + O_2 + M$	$7.0 \cdot 10^{15}$	0	192.1
(13)	$H + HO_2 \rightarrow OH + OH$	$1.5 \cdot 10^{14}$	0	4.2
(14)	$H + HO_2 \rightarrow H_2 + O_2$	$2.5 \cdot 10^{13}$	0	2.9
(15)	$O + HO_2 \rightarrow OH + O_2$	$2.0 \cdot 10^{13}$	0	0
(16)	$OH + HO_2 \rightarrow H_2O + O_2$	$2.0 \cdot 10^{13}$	0	0

$$k = A(T/K)^b \exp(-E/RT), \quad n = \text{reaction order}$$

$$[M] = [H_2] + 0.4[O_2] + 0.4[N_2] + 6.5[H_2O]$$

Table 1: Mechanism of H_2 oxidation (reactions forming and consuming H_2O_2 neglected); for reference see [4].

2. Formulation of Test Problem B

The problem under consideration is a stoichiometric laminar flat H_2-air flame at atmospheric pressure. The fixed reaction scheme is listed in Table 2 [7,8].

No.	Reaction	A $(cm^3/mol)^{n-1}s^{-1}$	b	E kJ/mol
(1)	H_2 + OH → H_2O + H	$2.2 \cdot 10^{13}$	0	21.5
(2)	O_2 + H → O + OH	$2.2 \cdot 10^{14}$	0	70.3
(3)	H_2 + O → H + OH	$1.8 \cdot 10^{10}$	1	37.2
(4)	OH + OH → H_2O + O	$6.3 \cdot 10^{12}$	0	4.6
(5)	H + H+M → H_2 + M	$6.4 \cdot 10^{17}$	-1	0
(6)	OH + O+M → HO_2 + M	$5.0 \cdot 10^{16}$	0	0
(7)	O_2 + H+M → HO_2 + M	$1.5 \cdot 10^{15}$	0	7.9
(8)	HO_2 + H → OH + OH	$2.5 \cdot 10^{14}$	0	7.9
(9)	HO_2 + H → H_2 + O_2	$2.5 \cdot 10^{13}$	0	2.9
(10)	OH + HO_2 → H_2O + O_2	$1.5 \cdot 10^{13}$	0	0
(11)	HO_2 + O → O_2 + OH	$6.3 \cdot 10^{13}$	0	2.9

$k = A(T/K)^b \exp(-E/RT)$, n = reaction order

[M] = total concentration

Table 2: Mechanism of H_2 oxidation to be used for the solution of test problem B [7,8].

The use of reverse reactions of reactions (1) to (4) is required; the data may be taken either from [7,8] or from equilibrium constants (see e.g. [9]). Reverse reactions of reactions (5) to (11) are optional to allow for the use of programs including reverse reactions in any case.

This reaction scheme is suggested for convenience irrespective of the accuracy of the data and of the disagreement of calculated and experimental flame velocity. (The main deficiency of this mechanism is the use of a uniform collision efficiency for all species in reactions (5) to (7) ; the inclusion of reaction (6) instead of H + OH + M → H_2O + M is the result of a misprint.)

The exact formulation of the governing equations can be chosen by the authors solving the test problem and shall be documented. Nevertheless, the problem demands the solution of one-dimensional conservation equations for total mass m, enthalpy H, and masses m_i of species i similar to the following equations (see[10-12]):

$$\text{m:} \quad \frac{\partial \varrho}{\partial t} = - \frac{\partial(\varrho v)}{\partial z} \tag{1}$$

$$\text{H:} \quad \varrho \frac{\partial T}{\partial t} = - \varrho v \frac{\partial T}{\partial z} + j_H \frac{\partial T}{\partial z} + \frac{1}{c_p} \frac{\partial}{\partial z} \left(\lambda \frac{\partial T}{\partial z} \right) - \frac{\sum r_i h_i}{c_p} \tag{2}$$

$$\text{m}_i\text{:} \quad \varrho \frac{\partial w_i}{\partial t} = - \varrho v \frac{\partial w_i}{\partial z} - \frac{\partial j_i}{\partial z} + r_i \tag{3}$$

Convection Diffusion Heat conduction Reaction

with $$j_H = \frac{\sum c_{p,i}\, j_i}{c_p}$$

The mass fractions w_i and the temperature T are independent variables in this example (c_p = specific heat capacity, h = specific enthalpy, j_i = diffusion flux of species i, r = mass scale chemical rate of formation, t = time, v = flow velocity, z = cartesian space coordinate, λ = mixture heat conductivity, ϱ = density).

The boundary conditions at the cold side are fixed to be

$$T_{cold} = T_{unburnt} = 298.15 \text{ K}$$

$$w_{i,cold} = w_{i,unburnt} = (\text{stoichiometric } H_2\text{-air mixture})$$

The boundary conditions at the hot side and (if necessary) the initial conditions with respect to the time t can be chosen by the authors and shall be documented.

Transport properties and heat capacities, again, are left to the authors' choice. In addition to the flame problem, calculated values of the heat capacities, the species heat conductivities, and binary diffusion coefficients are requested for the following mixture (corresponding to the burnt gas state of the test flame):

$T = 2000$ K, $w(H_2) = 3.196 \cdot 10^{-3}$, $w(H) = 8.825 \cdot 10^{-3}$,

$w(O_2) = 2.443\cdot10^{-2}$, $w(O) = 3.894\cdot10^{-3}$, $w(OH) = 1.020\cdot10^{-2}$
$w(H_2O) = 0.2113$, $w(HO_2) = 1.043\cdot10^{-5}$, $w(N_2) = 0.7461$

The method of solution of the conservation equations (1) to (3) is left to the authors' choice. Requested results are flame velocity, maximum values of the mass fraction profiles of H, O, OH, and HO_2, and the enthalpy profile.

3. Participating Groups

The following eight groups have solved test problem B:

Group 1: G. Dixon-Lewis

Department of Fuel and Energy
The University of Leeds
Leeds LS2 9JT , England

Group 2: J.M. Heimerl

Ballistic Research Laboratory
Aberdeen Proving Ground, MA 21005, USA

Group 3: H. Jinno, S. Fukutani

Department of Industrial Chemistry
Kyoto University , Japan

Group 4: R.D. Reitz
Department of Mechanical and Aerospace Engineering
Princeton University
Princeton, NJ 08544, USA

Group 5: Mitchell D. Smooke

Sandia National Laboratories
Livermore, CA 94550, USA

Group 6: H.J. Thies, N. Peters

Institut für Allgemeine Mechanik
RWTH Aachen
Templergraben 64
5100 Aachen, W.Germany

Group 7: J. Warnatz

Institut für Physikalische Chemie
Technische Hochschule Darmstadt
Petersenstr. 20
6100 Darmstadt, W.Germany

Group 8: Elaine S. Oran

Laboratory for Computational Physics
Naval Research Laboratory
Washington D.C. 20375, USA

4. Description of the Numerical Methods Used by the Different Groups

There are six groups solving the time-dependent conservation equations (Eq. 1 to 3 in Section 2), partly by use of finite difference methods (Groups 1,2,4,6 and 7), partly by use of finite element methods (Group 2 ; more information is given in the discussion of test problem A). Group 5 , however, is solving the corresponding stationary equation (a description of this method is given in a separate paper).

On the one hand, fully time-dependent implicit (Groups 3 and 6) or stiffly stable explicit (Group 4) solutions have been used, taking into account coupling of the species conservation equations. On the other hand, simplified implicit relaxation methods, neglecting coupling in time of the species conservation equations (Groups 1,2 and 7) have been used. Details are given in Table 3, parts A and B.

For the description of the transport phenomena a large variety of different models and their combinations has been used(details in Table 3, part C). The problems connected with the proper choice of a transport model are discussed elsewhere (papers by Heimerl and by Warnatz).

In part D of Table 3, some information is given on the computers used and the corresponding calculation times. Direct comparison is not possible; nevertheless, these data give a representative survey of the computational effort necessary for detailed laminar flame calculations.

	Group 1	Group 2	Group 3	Group 4	Group 5	Group 6	Group 7	Group 8
	Dixon-Lewis	Heimerl	Jinno/Fukutani	Reitz	Smooke	Thies/Peters	Warnatz	Oran
A. METHOD								
Basic method	t-dependent uncoupled finite difference	t-dependent uncoupled method of lines	t-dependent ? method of lines	t-dependent coupled finite difference	stationary - finite difference	t-dependent coupled finite difference	t-dependent uncoupled finite difference	t-dependent coupled finite difference
Order (time/space)	1 / 1	1 / 4	?	1 / 2	- / 1	1 / 4	1 / 2	2 / 2
ϱv from	profiles[1)]	profiles[1)]	trial and error	v of reference point	additional dif.eq.	v of reference point	profiles[1)]	v of reference point
Hot boundary conditions	zero slope	zero slope	adiabatic T, w_i	open boundary	zero slope	open boundary	constant slope	open boundary
Choice of t step	$\propto$ to characteristic t	controled by error limit	increasing, trial and error	fixed, trial and error	-	controled by error limit	$\propto$ to characteristic t	?
B. GRID SYSTEM								
Number of points	33	50	40	51	53	40	36	≤ 100
Grid point distance	non-uniform (see [13])	non-uniform (see [6])	T(dT/dx)	uniform in transformed plane	T(dT/dx)	uniform	non-uniform (see [13])	smaller in flame front
$n(HO_2)$ [2)]	11	6	10	4	17	6	10	20
$n(T)$ [3)]	10	22	31	10	29	16	27	30

Table 3: Details of the methods used (to be continued)

	Group 1 Dixon-Lewis	Group 2 Heimerl	Group 3 Jinno/Fukutani	Group 4 Reitz	Group 5 Smooke	Group 6 Thies/Peters	Group 7 Warnatz	Group 8 Oran
C. TRANSPORT								
Mixture heat conductivity	complete	different models[4)]	simplified	simplified	simplified	simplified	simplified	simplified
Diffusion fluxes	complete	different models[4)]	simplified	complete	simplified	complete	simplified	complete
Thermal diffusion	complete	simplified	not included	not includes	simplified	simplified	simplified	not included
Source of potential parameters	Svehla [5)]	Heimerl/Warnatz [7)]	Perry [6)]	Bird et al. [8)]	Warnatz [7)]	Hirschfelder[9)]	Warnatz [7)]	Svehla [5)]
D. COMPUTER								
Type	Amdahl	Cyber 76	FACOM M-200	IBM 4341	CRAY-I	Cyber 175	IBM 370/168	Texas Instr. ASC
cpu time per t step	0.22 s	} 100 s total	0.025 s	3.5 s	} 55 s total	1.28 s	0.4 s	0.13 s
Number of t steps	~1000		~3000	~1000		~500	~1000	~250

1) assuming stationarity 2) with $x(HO_2)>10\%$ of maximum value 3) between 350 K and 0.9 T_{max}

5) ref. [14] 6) ref. [15] 7) ref. [15] 8) ref. [16] 9) [ref. [11]

Table 3: (continued)

	Group 1	Group 2	Group 3	Group 4	Group 5	Group 6	Group 7	Group 8
	Dixon-Lewis	Heimerl	Jinno/Fuku-tani	Reitz	Smooke	Thies/Pe-ters	Warnatz	Oran
v_u [m/s]	1.91	2.00	2.74	2.13	2.06	1.45	2.02[1)]	1.80
10^3 w(H,max.)	2.72	2.67	3.09	2.71	2.22	2.80	2.83	3.35[2)]
10^3 w(O,max.)	7.99	7.55	8.32	7.47	7.45	8.19	6.07	6.88[2)]
10^4 w(HO_2,max.)	6.00	7.93	11.4	7.89	6.26	6.25	7.72	31.2[2)]
h_{max} [J/g]	235	365	492	-	276	295	334	-
h_{min} [J/g]	-112	-94.9	-91.8	-	-104	-84.2	-106	-
$\lambda(N_2)$ [W/m·K]	0.114	0.112	0.111	0.105	0.112	0.130	0.110	0.114
$\mathcal{D}_{H-H_2}$ [cm²/s]	40.6	51.9	72.4	49.0	45.3	40.5	52.5	57.19
λ [W/m·K]	0.172	-0.186	0.292	0.188	0.179	0.195	0.181	0.130
$\mathcal{D}_{H-N_2}$ [cm²/s]	24.5	30.7	40.5	29.5	26.7	24.1	32.1	43.4

1) v_u = 2.16 m/s without thermal diffusion

2) calculated from mole fractions using results of Group 7

Table 4: Selected results of test problem B

5. Results

Some results of test problem B (flame velocity v_u, maximum values of the mass fractions w_i of H, O, OH, and HO_2, maximum and minimum value of the specific enthalpy h) are given in Table 4.

Furthermore, this table contains the heat conductivities and diffusion coefficients calculated at T = 2000 K for the mixture specified in connection with problem B to test the transport models used by the different groups.

The calculated flame velocities (first row in Table 4) range from 1.45 m/s to 2.74 m/s (the expected result is near 2.0 m/s, see Section 1). The computed maximum mass fractions of the radical species H,O, OH, and HO_2 (second to fifth row) show amazingly good agreement, whereas there are discrepancies between the maximum and the minimum values of the specific enthalpy h calculated by the different groups (sixth and seventh row in Table 4).

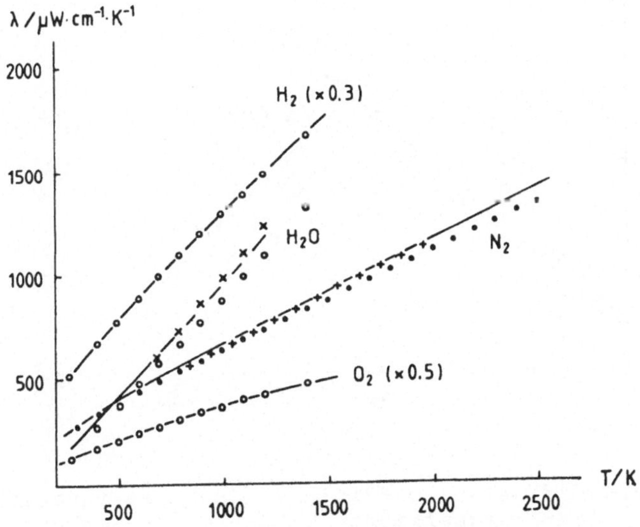

Fig. 2: Heat conductivities for stable species in the H_2-O_2-N_2 system. Points: experiments by different workers (for reference see [3]); lines: calculations using an extended Eucken correction (see paper by Warnatz).

The discrepancies in the flame velocity v_u and the enthalpy values h can be caused by deficiencies of the numerical scheme or by deficiencies of the transport model used. For clarifications, calculated transport properties at 2000 K ($\lambda(N_2)$ and $\mathcal{D}_{H-H_2}$, see second part of Table 4) can be compared with experimental data available for these two quantities.

In Fig. 2 the experimental heat conductivity $\lambda(N_2)$ is shown to be 0.113 W/m·K. Thus, there is sufficient agreement with the calculated values of all groups. (The relatively large value of Group 6 cannot be the reason for the low flame velocity, see below.)

The experimental binary diffusion coefficient $\mathcal{D}_{H-H_2}$ at T = 2000 K is about 55 cm^2/s, as shown in Fig. 3.

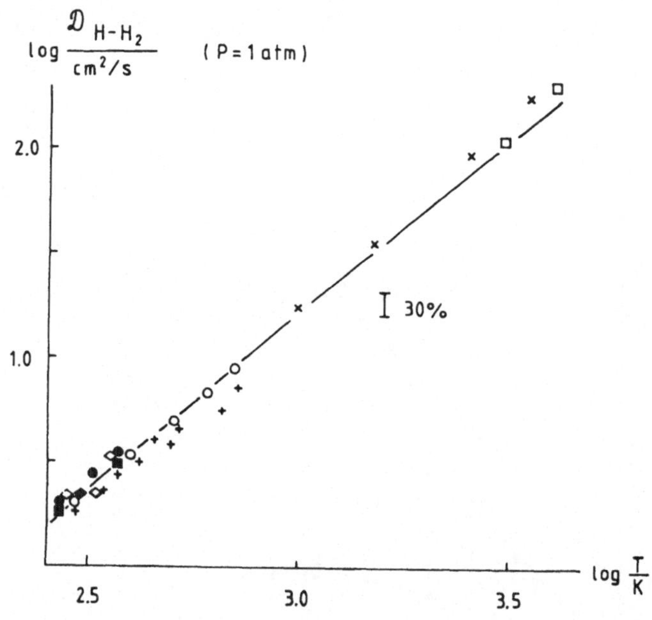

Fig. 3: Binary diffusion coefficient $\mathcal{D}_{H-H_2}$ as function of the temperature. Points: experiments by different workers (for reference see [17]); lines: calculated with $\sigma(H) = 2.05 \cdot 10^{-10}$m, $\varepsilon/k(H) = 145$ K, $\sigma(H_2) = 2.92 \cdot 10^{-10}$ m, $\varepsilon/k(H_2) = 38$ K (see paper by Warnatz)

This time, there are considerable deviations of the calculated values listed in Table 4, which can influence the computed flame velocity. On the other hand, the relatively large uncertainty of the experimental value of $\mathcal{D}(H\text{-}H_2)$ must be taken into consideration (see error bar in Fig. 3).

A reliable separation of the influences of deficiencies of the numerical scheme and the transport model is possible in the following way:

Simplified flame theory gives the following expression for the flame velocity [18]:

$$v = \sqrt{\frac{a}{\tau} \cdot F} \qquad \text{with} \qquad a = \frac{\lambda}{\varrho \cdot c_p}$$

Here τ is the mean reaction time (for the one-step reaction assumed) , and F is a function of the reaction type and the dimensionless burnt gas temperature with $F \approx 1$. τ is nearly equal in the calculations of all groups because of the common mechanism used. Thus, the quantity $\tau = \lambda / v_u^2 \cdot \varrho \cdot c_p$ must be similar for all groups, independent of the transport model used. If this quantity has different values, different flame velocities are caused by the numerical scheme.

This comparison is made in Table 5. It is evident that the low flame velocity calculated by Group 6 is caused by the numerical scheme, probably by insufficient efficiency of the equidistant grid point system used(N. Peters, private communication).

Group	v_u [m/s]	$\lambda / v_u^2 = \varrho \cdot c_p \cdot \tau$
(1) Dixon-Lewis	1.91	0.047
(2) Heimerl	2.00	0.047
(3) Jinno/Fukutani	2.74	0.039
(4) Reitz	2.13	0.041
(5) Smooke	2.06	0.042
(6) Thies/Peters	1.45	0.093
(7) Warnatz	2.02	0.039
(8) Oran	1.80	0.040

Table 5: Test of influence of the numerical scheme on calculated flame velocities v_u.

On the other hand, a test of the consistency of the transport model is given by the calculation of Lewis numbers $Le = \mathcal{D} \cdot \varrho \cdot c_p / \lambda$. The result is given in Table 6 with $\mathcal{D}(H-N_2)$ taken to be a representative diffusion coefficient for the non-thermal part of flame propagation.

Group	$\mathcal{D}(H-N_2)/\lambda = Le/\varrho \cdot c_p$
(1) Dixon-Lewis	142
(2) Heimerl	165
(3) Jinno-Fukutani	139
(4) Reitz	157
(5) Smooke	149
(6) Thies/Peters	124
(7) Warnatz	177
(8) Oran	334

Table 6: Lewis number (see text)

6. Conclusions

There are two substantial results of treating test problem B which are worth mentioning here:

(1) Numerics of laminar flame front calculations: The proper choice of a coordinate system and the construction of sophisticated grid point systems are precondition of reliable numerical results. With suitable gridding the calculation of laminar flame fronts is possible with use of about 25 grid points (see Fig.4; similar results are given in the paper by Smooke).

(2) Physics of laminar flame front calculations: The methods used for the simulation of the physical phenomena (in special for the calculation of transport properties) in flame front computations are well developed now, as can be shown by the agreement of results of a number of workers using different transport models. Flame front calculations seem to be reliable enough to allow the attack of the actual eminent problems connected with flame chemistry.

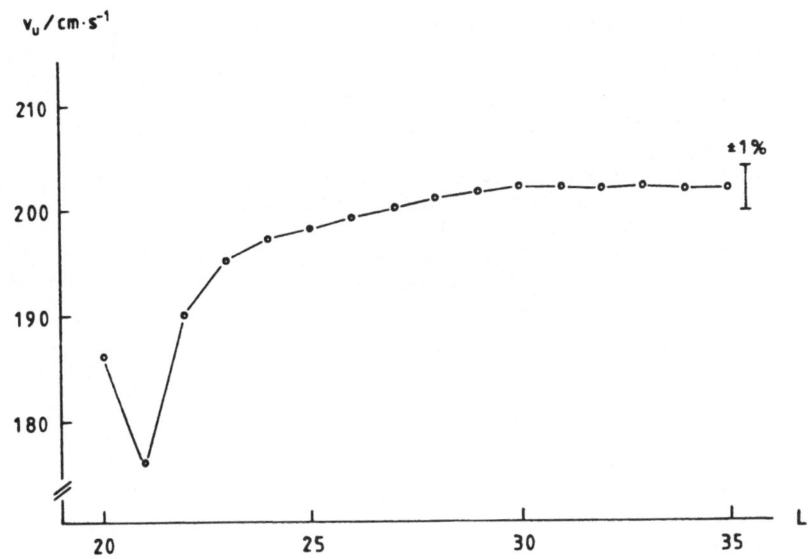

Fig. 4: Influence of the grid point number L on the flame velocity v_u of the flame of test problem B. Grid points are concentrated in the flame front (see Table 3).

References

[1] M.A. Cherian, P. Rhodes, R.J. Simpson, G.Dixon-Lewis, Eighteenth Symposium (International) on Combustion, p. 385. The Combustion Institute, Pittsburgh (1981)

[2] J. Warnatz, Comb.Sci.Technol. 26, 203 (1981)

[3] J. Warnatz, Ber.Bunsenges.Phys.Chem. 82, 643(1978)

[4] J. Warnatz, in: W.C. Gardiner, Flame Chemistry. Springer, New York (in press)

[5] G. Dixon-Lewis, Philos.Trans.Roy.Soc. 292, 45 (1979)

[6] J. Heimerl, A Contribution to the "Flat Flame Olympics": Test Problem B. Internal Report, Ballistic Research Laboratory, Aberdeen Proving Ground (1981)

[7] D.L. Baulch, D.D. Drysdale, D.G. Horne, A.C. Lloyd, Evaluated Kinetic Data for High Temperature Reactions, Vol. 1. Butterworths, London (1972)

[8] D.L. Baulch, D.D. Drysdale, J. Duxbury, S. Grant, Evaluated Kinetic Data for High Temperature Reactions, Vol. 3. Butterworths, London (1976)

[9] JANAF Thermochemical Tables. U.S. Department of Commerce, Washington D.C. (1971)

[10] J. Warnatz, Ber.Bunsenges.Phys.Chem. 82, 193 (1978)

[11] J.O. Hirschfelder, C.F. Curtiss, R.B. Bird, Molecular Theory of Gases and Liquids. Wiley, New York (1954)

[12] G. Dixon-Lewis, Proc.Roy.Soc. A 298, 495 (1967)

[13] D.B. Spalding, P.L. Stephenson, Proc.Roy.Soc. A 324, 315 (1971)
[14] R.A. Svehla, NASA Tech.Rept. R-132 (1962)
[15] J. Warnatz, Eighteenth Symposium (International) on Combustion, p. 369. The Combustion Institute, Pittsburgh (1981)
[16] R.B. Bird, W.S. Stewart, E.N. Lightfoot, Transport Phenomena. Wiley, New Yirk (1960)
[17] J. Warnatz, Berechnung der Flammengeschwindigkeit und der Struktur von laminaren flachen Flammen. Habilitationsschrift, Darmstadt (1977)
[18] D.A. Frank-Kamenetskii, Diffusion and Heat Exchange in Chemical Kinetics. Princeton University Press, Princeton (1955)

ON THE USE OF ADAPTIVE GRIDS IN NUMERICALLY CALCULATING ADIABATIC FLAME SPEEDS*

Mitchell D. Smooke
Applied Mathematics Division

James A. Miller
Combustion Chemistry Division

Robert J. Kee
Applied Mathematics Division

SANDIA NATIONAL LABORATORIES
Livermore, CA USA

ABSTRACT

We investigate the effects of using adaptive and equi-spaced grids in an adiabatic flame speed calculation.

1. Introduction

In this paper we want to illustrate the effects of using adaptive and equi-spaced grids in the numerical calculation of adiabatic flame speeds. The results we present are not designed to be a definitive statement as to how one should go about placing mesh nodes in the flame zone; they merely point out that accurate placement of grid points in regions where the dependent solution components exhibit rapid variation can lead to a significant reduction in the number of subintervals needed to obtain accurate flame speeds than if equi-spaced mesh intervals had been used. As a result, the overall cost of a flame speed calculation can be substantially reduced.

*Prepared by Sandia National Laboratory, Livermore, CA, 94550 for the United States Department of Energy.

2. Solution Method

We apply a global finite difference method to the pre-mixed flame equations. The calculation is initially performed on a coarse grid (usually 3-5 subintervals). The diffusion terms are approximated by centered difference expressions and the convective terms are upwinded. The resulting nonlinear difference equations are solved using a damped-modified Newton method. Once a converged numerical solution is obtained on an initial coarse grid, mesh points are inserted in regions where the dependent solution components exhibit high gradient and high curvature behavior (see [1] for details and also these proceedings).

In calculating adiabatic flame speeds (u) by steady-state methods the flow rate $\dot{M} = \rho u$, where ρ denotes the density of the mixture, must be determined such that the governing equations and boundary conditions have a solution (trivial or not). To solve this type of generalized eigenvalue problem we introduce a trivial differential equation ($d\dot{M}/dx=0$) for the flow rate along with an additional boundary condition. We then apply the damped-modified Newton method discussed in [1] to this augmented system. The extra boundary condition is chosen so that the temperature is specified at an interior point of the integration interval.

3. Numerical Results

We apply the steady-state solution method discussed above to calculate the adiabatic flame speed for a one atmosphere stoichiometric hydrogen-air flame (discussed in these proceedings). The flame model used in the calculations is discussed in [1-2]. The calculations are performed with equi-spaced and adaptively determined grids. The adaptive meshes are determined by seeking to equi-distribute changes in the components of the solution as well as changes in the gradients of the components of the solution between consecutive grid points [1]. For completeness, we list the reaction mechanism used in the calculation in Table I.

TABLE I

HYDROGEN-AIR REACTION MECHANISM

$$k_j^f = A_j^f \, T^{\beta_j^f} \, e^{-E_j^f/RT} \quad \text{(units are moles, cm}^3\text{, sec)}$$

j	Reaction	A_j^f	β_j^f	E_j^f
1	$H_2 + OH \rightleftarrows H_2O + H$	2.2×10^{13}	0.0	5180.0
2	$H + O_2 \rightleftarrows OH + O$	2.2×10^{14}	0.0	16900.0
3	$O + H_2 \rightleftarrows OH + H$	1.8×10^{10}	1.0	8960.0
4	$OH + OH \rightleftarrows O + H_2O$	6.3×10^{12}	0.0	1100.0
5	$H + H + M \rightleftarrows H_2 + M$	6.4×10^{17}	-1.0	0.0
6	$OH + H + M \rightleftarrows H_2O + M$	5.0×10^{16}	0.0	0.0
7	$O_2 + H + M \rightleftarrows HO_2 + M$	1.5×10^{15}	0.0	-1000.0
8	$H + HO_2 \rightleftarrows H_2 + O_2$	2.5×10^{13}	0.0	700.0
9	$H + HO_2 \rightleftarrows OH + OH$	2.5×10^{14}	0.0	1900.0
10	$OH + HO_2 \rightleftarrows H_2O + O_2$	1.5×10^{13}	0.0	0.0
11	$HO_2 + O \rightleftarrows O_2 + OH$	6.3×10^{13}	0.0	700.0

In the first set of calculations we determined flame speeds on grids consisting of 20, 40, 80, 160, 320 and 640 equi-spaced points. The results of the calculations are listed in Table II.

TABLE II

HYDROGEN-AIR FLAME SPEEDS (cm/sec)
(Equi-spaced Grids)

No. of Grid Points	20	40	80	160	320	640
Flame Speed (u) =	445	289	244	211	193	184

The second set of calculations was performed using the adaptive grid procedure outlined above. The flame speeds were determined on grids consisting of 20, 30, 40, 50 and 60 adaptively determined points. The results of the calculations are listed in Table III.

TABLE III
HYDROGEN-AIR FLAME SPEEDS (cm/sec)
(Adaptive Grid)

No. of Grid Points	20	30	40	50	60
Flame Speed (u) =	248	212	185	181	181

Several points are worth discussing. First, for both the equi-spaced and the adaptively determined grids, we see that as the number of mesh intervals increases, the flame speeds decrease. Second, the sequence of flame velocities obtained in the adaptive calculations approach a limiting value with only 40-50 grid points, while flame velocities obtained in the equi-spaced calculations are still changing by almost 15 percent as we go from 80 to 160 grid points. In fact, it was not until we used 640 equi-spaced points that we were able to obtain flame speeds that were within 2 percent of the flame speed calculated on the 50 point adaptive grid.

The decrease in flame speeds as we go to finer and finer grids is related to the effect of numerical diffusion and/or conduction. Recall that we have used upwind difference expressions for the convective terms in the flame equations. For grids in which the mesh spacing is denoted by h_j, we are, in effect through first order in h_j, solving the premixed flame equations with extra diffusion and conduction terms the sizes of which are proportional to $\rho u h_j/2$. Hence, the effective mass and thermal diffusivities are larger on coarser grids than on fine ones. When we combine this result with the fact that, for flames in which the velocity of reaction is appreciable at temperatures close to the temperature of the burned mixture, the adiabatic flame speed $u \sim \sqrt{\lambda / C_p}$ [3], it is apparent why we obtain higher flame speeds on coarse as opposed to fine grids.

If centered differences had been used the size of the numerical diffusion and conduction terms would be proportional to $\rho u(h_j - h_{j-1})/2$. In such a calculation we do not expect to obtain flame speeds as high as those obtained with upwind approximations on the coarse grids. In addition for the finer grids, the effect of numerical diffusion and/or conduction becomes less important and we expect similar results for both the upwind and centered calculations.

In the adaptive calculation with 50 grid points the ratio of the integration interval to the minimum mesh interval -- L/h_{min} -- is equal to 625. This quantity gives a representative value for the number of mesh intervals one should use in an equi-spaced calculation if results comparable to the adaptive calculation are to be obtained. The size of this number helps to explain why we obtained flame speeds that were within 2 percent of the 50 point adaptive calculation only after 640 grid points were used in the equi-spaced calculation.

The 50 point adaptive calculation took 45 seconds of CPU time on a Cray-I while the 640 point equi-spaced calculation took 327 seconds. Hence, for this problem, a savings of about a factor of seven results in going from equi-spaced to adaptive grids.

4. Remarks

Numerical diffusion can significantly affect calculated adiabatic flame speeds on coarse grids. The effect is more pronounced if, for a fixed number of grid points, one uses an equi-spaced mesh as opposed to an adaptively determined mesh. In addition, by placing grid points in regions in which the dependent solution components exhibit rapid variation, one can dramatically reduce the number of mesh intervals (and hence the cost of the calculation) required to obtain accurate flame speeds than if an equi-spaced mesh had been employed.

REFERENCES

1. M. D. Smooke, Sandia National Laboratories Report SAND81-8040, (1982).

2. J. A. Miller, R. E. Mitchell, M. D. Smooke and R. J. Kee, submitted to the 19th Symposium (International) on Combustion, (1982).

3. Y. B. Zeldovich and D.A. Frank-Kamenetsky, Dokl. Akad. Nauk. SSSR 19, (1938) 693.

RESULTS OF A STUDY OF SEVERAL TRANSPORT ALGORITHMS FOR PREMIXED, LAMINAR STEADY-STATE FLAMES

by

J.M. Heimerl and T.P. Coffee

Ballistic Research Laboratory
Maryland, USA

1. INTRODUCTION

Our model /1/ of a premixed, steady-state flame includes detailed elementary chemical reactions and requires as input not only the kinetics information (of our immediate interest), but also thermodynamic and transport data. Fortunately for the types of chemical species we are interested in, the thermodynamics input is by and large well defined /2/, /3/. In addition, while some transport coefficients are only well defined through low temperature (< 1000 K) measurements /4/, the theory is sufficiently developed to allow reasonable estimates to be made at higher temperatures /5/. A theory has been developed for multicomponent mixtures /5/ - /9/, but it is computationally cumbersome. To circumvent this, previous workers have generally employed some level of simplification /10/ - /20/.

In a recent paper /21/ we have addressed in some detail the question: which of the mathematical approximations to the multicomponent transport properties provides a desirable trade-off between precision and computational effort. Another way of phrasing this question is to ask what loss in precision of predicted flame speeds and profiles occurs as the mathematical approximations to the multicomponent, polyatomic transport expressions are made cruder.

We have approached this problem by actually computing the properties of the H_2-O_2-N_2 which has a set of well characterized input parameters. We fix these input parameters and vary the transport algorithms. The computed flame speeds and profiles are then compared. The numerical method is discussed in /22/, /23/, and the details of the input parameters and transport algorithms are discussed elsewhere /17/, /24/. It shall suffice here to simply enumerate the approximations used and briefly discuss the results.

2. TRANSPORT ALGORITHMS

We shall now outline five approximations to the multicomponent, polyatomic formalism, based on the theory of Wang Chang and Uhlenbeck /5/ - /9/. We start with the most accurate and progressively consider cruder approximations.

Method I

We can write for the heat flux

$$q = \sum_{i=1}^{N} \rho Y_i V_i h_i - \lambda_o T_x - \sum_{i=1}^{N} \frac{RTD_i^T}{M_i X_i} (X_i)_x \qquad (1)$$

and for the diffusion velocity, which enters into the mass flux, $\rho_i Y_i V_i$, we have

$$V_i = \frac{1}{X_i} \sum_{j=1}^{N} \frac{Y_j}{X_j} D_{ij} (X_j)_x - \frac{D_i^T}{\rho Y_i} (\ell n\ T)_x \quad . \qquad (2)$$

Expressions for D_{ij}, D_k^T and λ_o are computed by a formal expansion. Method I is the three term Sonine approximation to the formalism expressed in equations (1) and (2), /25/ gives a discussion of this approximation. The three term expansion requires the solutions of two matrix equations each involving a matrix of dimension 3NX3N, where N is the number of species. The elements of this matrix are complicated functions of the pressure, temperature, mole fractions, viscosities, binary diffusion coefficients, specific heats, collision numbers and collision integrals. For further details see /24/ or /25/.

Method II

The above formalism is quite complicated to work with, and so further simplifications are almost invariably made. We can simplify by taking only one term in the Sonine polynomial expansions. For diffusion this simplification can be rearranged /5/ to give the Stefan-Maxwell equations:

$$(X_i)_x = \sum_{j=1}^{N} X_i X_j (D_{ij})^{-1} (V_j - V_i) \quad , \qquad i = 1, \ldots, N \quad . \tag{3}$$

This set of equations is not independent, and the constraint

$$\sum_{i=1}^{N} Y_i V_i = 0 \tag{4}$$

must be used in place of one of the equations in (3). Then the diffusion velocities can be found by solving a set of N equations in N unknowns.

Thermal conductivity is also simplified in this manner. But the resulting expressions, λ_{mix}^{mon}, is valid only for a mixture of monatomic gases.

To define the heat conductivity for a mixture of polyatomic gases, we adopt Hirshfelder's Eucken-type relation /4/, /26/,

$$\lambda_{mix}^{poly} = \lambda_{mix}^{mon} + \sum_{i=1}^{N} \frac{\lambda i - \lambda i^{mon}}{1 + \sum\limits_{j \neq i} \frac{\mathcal{D}_{ii}}{\mathcal{D}_{ij}} \frac{X_j}{X_i}} \tag{5}$$

Method III

By making additional assumptions, the Stefan-Maxwell equations (3) can be further simplified. A common assumption is that all but the ith species move with the same velocity V. Then we find that

$$(X_i)_x = X_i(V - V_i) \sum_{j \neq i} \frac{X_j}{\mathcal{D}_{ij}} \quad . \tag{6}$$

Employing (4) we find

$$V = \frac{- Y_i V_i}{1 - Y_i} \tag{7}$$

which when substituted into (6) yields the formula recommended by Hirshfelder and Curtiss /27/

$$V_i = - \frac{(1 - Y_i)}{X_i \sum\limits_{j \neq i} \frac{X_j}{\mathcal{D}_{ij}}} (X_i)_x \quad . \tag{8}$$

Unfortunately, the expression in (8) does not in general satisfy the constraint of equation (4). One technique to satisfy this constraint is due to Oran and Boris /28/. They note that if a set of diffusion velocities V_i satisfy the Stefan-Maxwell equations (3), then so does the set $(V_i + V_c)$, where V_c is some constant. The value of V_c is chosen such that the constraint (4) is satisfied.

The heat conductivity formula employed at this level of approximation is taken from Mason and Saxena's /4/, /29/ simplification of (5), specifically

$$\lambda_o = \sum_{i=1}^{N} \frac{\lambda_i}{1 + \sum\limits_{j \neq i} \phi_{ij} (X_j / X_i)} \tag{9}$$

where

$$\phi_{ij} = \frac{1.065}{8^{1/2}} \left(1 + \frac{M_i}{M_j}\right)^{-1/2} \left\{1 + \left(\frac{\eta_i M_j}{\eta_j M_i}\right)^{1/2} \left(\frac{M_i}{M_j}\right)^{1/4}\right\}^2 . \qquad (10)$$

Here η_i and M_i are the viscosity and molecular weight of the ith species.

Method IV

Equation (8) or some analogous form has often been used to compute diffusion. However, the usual procedure has been to use (8) only to compute $V_1, \ldots, V_{N-1}$. Then V_N is computed from (4). This is less accurate than the Oran and Boris procedure, especially for V_N.

Also, an empirical formula for the thermal conductivity,

$$\lambda_o = 0.5 \left(\sum_{i=1}^{N} X_i \lambda_i + \left\{ \sum_{i=1}^{N} X_i / \lambda_i \right\}^{-1} \right) , \qquad (11)$$

is often used /4/, /30/.

Method IV is comprised of these common approximations.

Method V

In the case of binary mixture the Stefan-Maxwell equations (3) reduce to Fick's law. Specifically we have,

$$Y_1 V_1 = - \mathcal{D}_{12} (Y_1)_x \quad . \qquad (12)$$

A generalization of (12) can be made /31/, and yields

$$Y_i V_i = - D_{im} (Y_i)_x \quad , \qquad (13)$$

where

$$D_{im} = \frac{1 - X_i}{\sum_{j \neq i} \frac{X_j}{\mathcal{D}_{ij}}} \quad . \qquad (14)$$

Additional assumptions that $\rho^2 D_{im}$, $\rho\lambda$ and $(c_p)_{mix}$ are each independent of temperature are made and the procedure that permits an _a priori_ selection of these quantities has been discussed /24/.

3. RESULTS

Table 1 shows a summary of the five methods of computing the transport properties used in this paper. Five H_2-O_2-N_2 flames were selected and their initial conditions listed in Table 2. The total pressure is fixed at one atmosphere for all flames. The computed flame speed for each flame as a function of transport method is tabulated in Table 3 (The flame speeds for flame A are not corrected to 291 K, as has been done /11/. If this were done, the value AI for example would be 12.2 cm/s instead of 14.1 cm/s). The values of the flame speeds span a large range and for a given flame are essentially independent of the transport method. The largest difference between Method I, the most complete formulation of the transport, and any other method is 16% (compare Methods I and III, flame D).

Note that even our _a priori_ method of selecting constant transport coefficients, Method V, gives results that are quite close to the much more complex Method I.

Reproduction of flame speeds is a necessary but not sufficient condition to judge the relative effectiveness of the transport methods. We must also examine the species and temperature profiles of these flames. As an example we consider a profile that exhibits differences among the five methods that are as large as any observed. Figures 1 and 2 show the OH profiles for flame D. As can be seen these profiles are very similar. The other species profiles and the temperature profiles show at least this degree of similarity among the five profiles as in the example given.

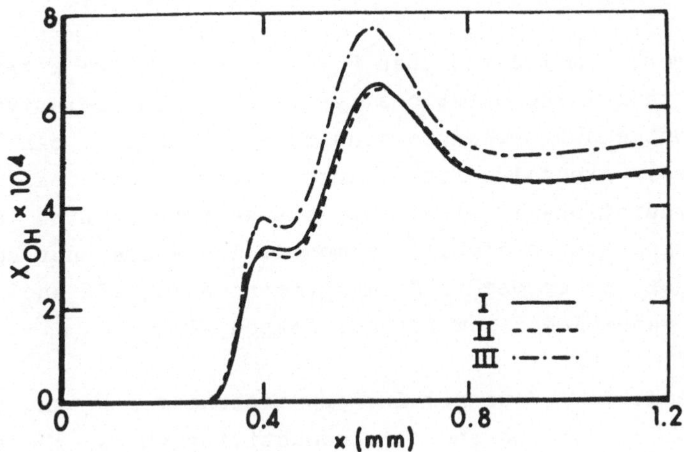

Figure 1: The OH Profile for Flame D; Transport Methods I, II and III .

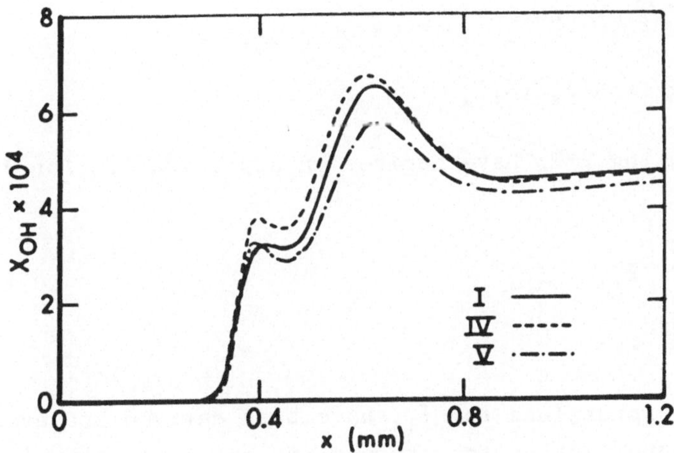

Figure 2: The OH Profile for Flame D; Transport Methods I, IV and V.

4. A NEW METHOD

For flames more complicated than H_2-O_2-N_2 we require a transport algorithm that is computationally efficient and relatively precise. The most exact procedure considered, Method I, can become prohibitively expensive for a large number of species. Method V is computationally efficient, but we feel caution is necessary in using such a simplified model. There does not seem to be a great deal to choose from among Methods II, III and IV and so we have assembled a new method, Method VI.

As with Methods II - V we require expressions for λ_o and for the V_i (= $\mathcal{V}_i + \mathcal{W}_i$). For the thermal conductivity we use the simplest formula (11), since the exact choice does not appear to be important /21/. For the molecular diffusion velocities, $\mathcal{V}_i$, we use the expression in equation (8).

For problems with light species and steep temperature gradients, the neglect of thermal diffusion is often as important as the differences between the computational methods. So we have generated a technique /21/ that approximates the thermal diffusion velocity, $\mathcal{W}_i$. We define

$$\mathcal{W}_i = k_{im} \mathcal{V}_i (\ln T)_x / (X_i)_x \quad . \tag{15}$$

Chapman and Cowling /32/ have derived an approximation for k_{im}, specifically

$$k_{im} = \sum_{\substack{j=1 \\ i \neq j}}^{N} k_{ij} \quad . \tag{16}$$

Theoretical expressions for k_{ij} have been derived and even in the first approximation the expressions are quite complicated /5/. For the special case of heavy isotopes, however, the first approximation simplifies to /5/

$$\ell_{ij} = \frac{15(2A^* + 5)(6C^* - 5)(M_i - M_j)}{2A^*(16A^* - 12B^* + 55)(M_i + M_j)} X_i X_j \; , \qquad (17)$$

where the starred quantities are ratios of collision integrals. Equation (17) reproduces H_2-N_2 thermal diffusion ratios to within 30%, and so we use it as a simple but reasonable approximation for ℓ_{ij}.

From Eq. (17) we can see that the influence of $\mathcal{W}_i$ will be the greater the larger the mass differences. Normally the thermal diffusion ratio does not exceed 0.1 and for the H_2-O_2-N_2 system, we have computed $\mathcal{W}_i$ only for the species H and H_2. The resulting diffusion velocities $V_i = \mathcal{V}_i + \mathcal{W}_i$ do not satisfy the constraint $\sum_{i=1}^{N} Y_i V_i = 0$. We use the Oran and Boris procedure discussed in Method III to obtain this condition.

This new method for approximating multispecies transport has been exercised in our test flames. Table 3 shows computed values for the flame speeds. The results are so close to those of Method I that we infer that some of the errors made in the approximations involved in Method VI are cancelling.

Figure 3 compares the OH profiles for flame D. In general, for the minor species, the accuracy of Method VI is comparable to Methods II or III. However, in almost every profile involving a major species (not shown), the accuracy was improved.

5. MISCELLANEOUS COMMENTS

As an aid for comparing efficiency among transport methods we have computed the so-called "grind-time" for each of the six transport methods and five flames discussed here. We take as the definition:

grind-time = CPU time/# time steps/# collocation points.

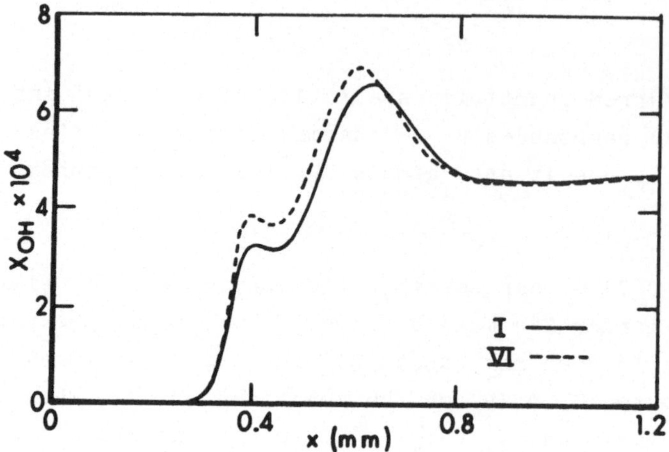

Figure 3: The OH Profile for Flame D; Transport Methods I and VI.

(We have used a relaxation technique employing the method of lines and the collocation points are those spatial locations at which the solution is found.) The results are given in Table 4.

Some caution must be exercised in the interpretation of Table 4. Rigorously we would require that all cases start from the same initially guessed profiles. Since this table was constructed after the fact /33/, this ideal was not met. Nevertheless, we offer the values in Table 4 as a guide to efficiency.

In the spirit of approximation exemplified in the discussion of Method V, the further assumption is sometimes made that all Lewis numbers ($Le_i = \lambda/c_p\ \rho D_{im}$) are equal to unity. Given values for $\rho\lambda$ and c_p, this assumption defines the diffusion coefficients and removes the requirement of supplying independent information for them. We tested how well this additional level of simplification fared by applying it in the code using flame C. The computed flame speed was about 40% lower than for Method I. While the major species profiles were still reasonably close to those of Method I, they were noticeably less accurate than Method V. The minor species profiles showed large differences: the peak concentrations of O, HO_2 and H_2O_2 were lower than those of Method I by about a factor of two.

We performed some checks on the stoichiometric methane-air flame. For this flame we used the input parameters listed by Tsatsaronis /13/. The 14 species kinetics scheme was used as listed and, as with the 8 species H_2-O_2-N_2 kinetics scheme used above /17/, we made no attempt to critically evaluate it. The flame speeds computed for Methods I, V and VI are in the ratio 1.00 to 0.88 to 0.94, respectively. The profiles for Method VI are more accurate than those produced by Method V. The grind-time for Methods I, V and VI are 124.4, 12.4 and 44.9 msec, respectively. These results follow the same trend as in the H_2-O_2-N_2 flames reported above.

6. DISCUSSION

The numerical results shown in Table 3 demonstrate that reliable results can be obtained for the H_2-O_2-N_2 system even for the case of our a priori determined constant transport method. Note that we cannot infer that transport is unimportant! The computed profiles and flame speeds can be sensitive to the choice of transport parameters selected. For the relative tests of the transport methods here, we have employed the same set of species viscosities, thermal conductivities and binary diffusion coefficients in all cases. We have demonstrated that the method used to generate the multicomponent, polyatomic transport coefficients is not critical for the H_2-O_2-N_2 flame, and does not appear to be critical for a stoichiometric methane-air flame. Since these flames are reasonably complex we infer that this result has a high probability of being valid for other flames.

Indeed, we conclude that gross errors detected in comparing the results of different models are more likely to be traceable to differences in input data rather than to the method of approximating multicomponent polyatomic transport properties.

In summary we find that the choice of a multicomponent transport algorithm is not critical.

TABLE 1: SUMMARY OF TRANSPORT METHODS

Method	Remarks
I	3 terms of Sonine polynomial expansion (3 N by 3 N matrix); only method that has non-zero thermal diffusion.
II	For diffusion, 1 term of Sonine polynomial expansion (N by N matrix). For thermal conductivity, Hirshfelder-Eucken method (N by N matrix), Eq. (5).
III	Diffusion velocities computed from simplified Stefan-Maxwell relation, Eq. (8). Each is adjusted by a common factor, V_c, so as to satisfy Eq. (4). Thermal conductivity from Mason and Saxena, Eqs. (9) and (10).
IV	Diffusion velocities computed from simplified Stefan-Maxwell relation, Eq. (8), for N-1 species. V_N is computed from Eq. (4). Empirical thermal conductivity formula, Eq. (11).
V	Diffusion velocities from generalized Fick's law, Eqs. (13) and (14). Empirical thermal conductivity formula, Eq. (11). In addition $\rho^2 D_{1m}$ = constant ; $\rho\lambda$ = constant ; c_p = constant ; constants are determined *a priori*.

TABLE 2: INITIAL TEMPERATURE AND MOLE FRACTIONS FOR THE FIVE FLAMES STUDIED

Flame	XH2	XO2	XN2	Tu
A	.1883	.0460	.7657	336
B	.2000	.1680	.6320	298
C	.5000	.1050	.3950	298
D	.9000	.1000	0	298
E	.6000	.4000	0	298

TABLE 3: FLAME SPEEDS CALCULATED USING THE FIVE TRANSPORT MODELS

Flame	I	II	III	IV	V	VI
A	14.1	14.6	14.9	14.9	16.0	14.0
B	98	101	102	103	96	97
C	292	300	310	308	291	292
D	378	379	438	402	348	402
E	892	922	971	969	847	894

TABLE 4: "GRIND-TIME" IN MSEC* FOR FLAMES AND TRANSPORT METHODS

Flame						
A	38.5	19.8	16.6	10.5	5.3	17.3
B	41.1	22.4	23.4	17.0	5.7	19.9
C	40.9	27.0	18.5	15.3	5.1	16.1
D	28.9	21.8	18.0	9.4	4.2	14.3
E	57.8	23.2	14.8	8.3	3.8	12.4
Average	41.4	22.8	18.3	12.1	4.8	16.0

* execution on the BRL CYBER 76 computer, single precision (64 bit word)

REFERENCES

/1/ J.M. Heimerl and T.P. Coffee, Combust. Flame, 39, 301-315 (1980).

/2/ G.S. Gordon and B.J. McBride, NASA-SP-273, 1971, Computer Program for Calculation of Complex Chemical Equilibrium Compositions, Rocket Performance, Incident and Reflected Shocks, and Chapman-Jouquet Detonations, (1976 program version).

/3/ D.R. Stull and H. Prophet, JANNAF Thermochemical Tables, 2nd Edition, NSRDS-NBS-37, June 1971.

/4/ Y.S. Touloukian, P.E. Liley and S.C. Saxena, Thermophysical Properties of Matter, Vol. 3, Thermal Conductivity (Nonmetallic Liquids and Gases), IFI/Plenum, NY-Washington (1970).

/5/ J.O. Hirshfelder, C.F. Curtis and R.B. Bird, "Molecular Theory of Gases and Liquids", 2nd printing, corrected, with notes, John Wiley and Sons, NY, 1960.

/6/ C.S. Wang Chang, G.E. Uhlenbeck and J. deBoer, Studies in Statistical Mechanics, Vol. 2, John Wiley and Sons, NY, 1964.

/7/ L. Monchick, K.S. Yun and E.A. Mason, "Formal Kinetic Theory of Transport Phenomena in Polyatomic Gas Mixtures", J. Chem. Phys. 39, 654-669 (1963).

/8/ L. Monchick, A.N.G. Pereira and E.A. Mason, "Heat Conductivity in Polyatomic and Polar Gases and Gas Mixtures", J. Chem Phys. 42, 3241-3256 (1965).

/9/ L. Monchick, R.J. Munn and E.A. Mason, "Thermal Diffusion in Polyatomic Gases: A Generalized Stefan-Maxwell Diffusion Equation", J. Chem. Phys. 45, 3051-3058 (1966).

/10/ J. Warnatz, Ber. Bunsenges. Phys. Chem., 1978, 82, 193-200, Calculation of the Structure of Laminar Flat Flames I: Flame Velocity of Freely Propagating Ozone Decomposition Flames.

/11/ G. Dixon-Lewis, et al., "Flame Structure and Flame Reaction Kinetics", Proc. R. Soc., London A 317, 235-263 (1970); A 331, 571-584 (1973); and A 346, 261-278 (1975).

/12/ G. Dixon-Lewis, "Kinetic Mechanism, Structure and Properties of Premixed Flames in Hydrogen-Oxygen-Nitrogen Mixtures", Phil. Trans. R. Soc., London, 292, 45-99 (1979).

/13/ G. Tsatsaronis, "Prediction of Propagating Laminar Flames in Methane, Oxygen, Nitrogen Mixtures", Combust. and Flame 33, 217-239 (1978).

/14/ D.B. Spalding and P.L. Stephenson, "Laminar Flame Propagation in Hydrogen and Bromine Mixtures", Proc. R. Soc., London A 324, 315-337 (1971).

/15/ P.L. Stephenson and R.G. Taylor, "Laminar Flame Propagation in Hydrogen, Oxygen, Nitrogen Mixtures", Combust. and Flame 20, 231-244 (1973).

/16/ L.D. Smoot, W.C. Hecker and G.A. Williams, "Prediction of Propagating Methane-Air Flames", Combust. and Flame 26, 323-342 (1976).

/17/ J. Warnatz, "Calculation of the Structure of Laminar Flat Flames II; Flame Velocity and Structure of Freely Propagating Hydrogen-Oxygen and Hydrogen-Air Flames", Ber. Bunsenges. Phys. Chem. 82, 643-649 (1978).

/18/ D.B. Spalding, P.L. Stephenson and R.G. Taylor, "A Calculation Procedure for the Prediction of Laminar Flame Speeds", Combust. and Flame 17, 55-64 (1971).

/19/ L. Bledjian, "Computation of Time-Dependent Laminar Flame Structure", Combust. and Flame 20, 5-17 (1973).

/20/ E. Cramarossa and G. Dixon-Lewis, "Ozone Decomposition in Relation to the Problem of the Existance of Steady-State Flames", Combust. and Flame 16, 243-251 (1971).

/21/ T.P. Coffee and J.M. Heimerl, "Transport Algorithms for Premixed, Laminar Steady-State Flames", Combustion and Flame 43, 273-289 (1981).

/22/ T.P. Coffee and J.M. Heimerl, "A Method for Computing the Flame Speed for a Laminar, Premixed, One-Dimensional Flame", BRL Technical Report, ARBRL-TR-02212, Jan. 1980.

/23/ T.P. Coffee, "A Computer Code for the Solution of the Equations Governing a Laminar, Premixed, One-Dimensional Flame", BRL Memorandum Report, in press.

/24/ T.P. Coffee and J.M. Heimerl, "Transport Algorithms for Premixed Laminar Steady-State Flames", BRL Technical Report, ARBRL-TR-02302, March 1981. Also see Ref. 21.

/25/ G. Dixon-Lewis, "Flame Structure and Flame Reaction Kinetics II. Transport Phenomena in Multicomponent Systems", Proc. R. Soc., London A 307, 111-135 (1968).

/26/ J.O. Hirshfelder, "Heat Conductivity in Polyatomic, Electronically Excited, or Chemical Reacting Mixtures III", Sixth International Combustion Symposium, Reinhold Publishing Corporation, NY, 351-366 (1957).

/27/ J.O. Hirshfelder and C.F. Curtiss, "Theory of Propagation of Flames Part I: General Equations", Third International

Combustion Symposium, Williams and Wilkins Co., Baltimore, 121-127 (1949).

/28/ E.S. Oran and J.P. Boris, "Detailed Modeling of Combustion Systems", Proj. Energy Combust. Sci. 7, 1-71 (1981).

/29/ E.A. Mason and S.C. Saxena, "Approximate Formula for the Thermal Conductivity of Gas Mixtures", Phys. Fluids 1, 361-369 (1958).

/30/ J.H. Burgoyne and F. Weinberg, "A Method of Analysis of a Plane Combustion Wave", Fourth Symposium on Combustion, Williams and Wilkins Co., Baltimore, 294-302 (1953).

/31/ R.B. Bird, W.S. Stewart and E.N. Lightfoot, "Transport Phenomena", John Wiley and Sons, NY, 1960.

/32/ S. Chapman and T.G. Cowling, The Mathematical Theory of Non-Uniform Gases, third edition, Cambridge University Press (1970).

/33/ Suggestion of R.D. Reitz, Princeton University, 1981.

Influence of Transport Models and Boundary Conditions on Flame Structure

Jürgen Warnatz
Institut für Physikalische Chemie der Technischen Hochschule
6100 Darmstadt, W. Germany

1. Introduction

1.1 Some General Remarks

Calculations of laminar flame fronts demand

(1) input data on thermodynamic properties,
(2) specifications of a transport model and input data for the evaluation of transport properties,
(3) input data on the chemical reactions occuring,
(4) specification of proper boundary conditions, and
(5) an appropriate numerical scheme for the integration of the conservation equations.

Point (1) does not lead to serious difficulties and does not need further consideration here. Because of the complicated problems involved, points (3) and (5) have attracted much attention in the last years (see chapter on test problem B). Because of this overwhelming interest and the rapid progress in flame chemistry and in the development of stiff-stable integration methods, problems(2) and (4) have been disregarded in the recent years, though they have been very often discussed in the literature of the decades before (for reference see [1]). This shall be the reason to consider some details of these problems in the following chapters 2 and 3.

1.2 One-Dimensional Conservation Equations and Solution Method

Conservation of enthalpy and of mass of species i leads to the time-dependent equations [2-4]

H: $$\rho \frac{\partial T}{\partial t} = -\rho v \frac{\partial T}{\partial z} + j_H \frac{\partial T}{\partial z} + \frac{1}{c_p}\frac{\partial}{\partial z}\left(\lambda \frac{\partial T}{\partial z}\right) - \frac{\Sigma r_i h_i}{c_p} \quad (1)$$

m_i: $$\rho \frac{\partial w_i}{\partial t} = -\rho v \frac{\partial w_i}{\partial z} - \frac{\partial j_i}{\partial z} + r_i \quad (2)$$

Convection Diffusion Heat conduction Reaction

where the diffusion fluxes j_i and the mean diffusion flux j_H are given by

$$j_H = \frac{\sum c_{p,i} j_i}{c_p} \quad ; \quad j_i = -D_{i,M}\, \rho \frac{\partial w_i}{\partial z} \quad (3)$$

(c_p = specific heat capacity, h = specific enthalpy, r = mass scale chemical rate of formation, t = time, T = temperature, v = flow velocity, w = mass fraction, z = cartesian space coordinate, λ = mixture heat conductivity, ρ = density).

A simplified transport model given by Eq.(4) and (5)

$$D_{i,M} = \sum_{j \neq i} \frac{1-w_i}{x_j / \mathcal{D}_{ij}} \quad ; \quad \sum_i j_i = 0 \quad (4)$$

$$\lambda_M = \frac{1}{2}\left(\sum_i x_i \lambda_i + \frac{1}{\sum_i x_i / \lambda_i}\right) \quad (5)$$

(M = mixture, x = mole fraction) is used for the calculations presented in chapter 3, because comparison with multicomponent transport models (discussion in chapter 2) results in relatively small errors. The binary diffusion coefficients $\mathcal{D}_{ij}$ and the pure species heat conductivities λ_i are calculated from angle-independent Stockmaier potential parameters, including an extended Eucken correction for the heat conductivity (see chapter 2).

Due to stiffness of the system of differential equations (1)/(2), an implicit finite difference method is chosen for solution [2,4]. This relaxation method starts with arbitrary profiles of temperature T and mass fractions w_i at time zero. With the aid of a grid point system the derivatives are replaced by finite difference expressions assuming a parabolic approach between three neighbouring grid points in each case. This procedure reduces the given problem to the solution of a tridiagonal linear equation system, if at the edges of the grid point system the values of temperature T and mass fractions w_i are specified by means of proper boundary conditions (see chapter 3).

The reaction mechanism used is given in [3]. It is very similar to that presented in the introduction to test problem b; therefore further discussion is unnecessary here.

2. Influence of Transport Models on Flame Structure

2.2 Multicomponent Transport Models

The multicomponent diffusion fluxes are given by [1]

$$j_i = \frac{M_i c^2}{\varrho} \sum M_j D_{ij} \frac{\partial x_j}{\partial z} - D_{T,i} \frac{\partial \ln T}{\partial z} \tag{6}$$

with multicomponent diffusion coefficients defined by [1,5]

$$D_{ij} = x_i \frac{16\,T}{25\,P} \frac{\bar{M}}{M_j} (K_{ij} - K_{ii}) \tag{7}$$

$$L_{ij} = \frac{16\,T}{25\,P} \sum_k \frac{x_k}{M_i \mathcal{D}_{ik}} [M_j x_j (1-\delta_{ik}) - M_i x_i (\delta_{ij} - \delta_{ik})] \tag{7a}$$

(c = molar concentration, M = molar mass, P = pressure, δ = Kronecker symbol; the K_{ij} are elements of the inverse of the matrix with the elements L_{ij}.)

A multicomponent formulation of the mixture heat comductivity is given by Eq. (8) to (10). Some difficulties are connected with the calculation of that part of the pure species thermal conductivity caused by diffusion of internal energy of molecules (Eucken correction, see below).

$$\lambda = \lambda^{(trans)} + \lambda^{(int)} \tag{8}$$

$$\lambda^{(int)} = \sum_i \frac{x_i/\mathcal{D}_{ii}}{\sum_j x_j/\mathcal{D}_{ij}} (\lambda_i - \lambda_i^{(trans)}) \tag{9}$$

$$\lambda^{(trans)} = 4 \begin{vmatrix} L_{11} & \cdots & L_{1n} & x_1 \\ \vdots & \ddots & \vdots & \vdots \\ L_{n1} & \cdots & L_{nn} & x_n \\ x_1 & \cdots & x_n & 0 \end{vmatrix} \Big/ \begin{vmatrix} L_{11} & \cdots & L_{1n} \\ \vdots & \ddots & \vdots \\ L_{n1} & \cdots & L_{nn} \end{vmatrix} \tag{10}$$

$$L_{ij} = \frac{16T \; x_i x_j \; M_i \; M_j}{25P \; \mathcal{D}_{ij} \; (M_i + M_j)^2} \left(\frac{55}{4} - 3 \, B^*_{ij} - 4 \, A^*_{ij} \right) \tag{10a}$$

$$L_{ii} = - \frac{4 \, x_i^2}{\lambda^{(trans)}} - \frac{16 \, T}{25 \, P} \sum_{k \neq i} S_{ik} \tag{10b}$$

$$S_{ik} = \frac{x_i \, x_k \left(\frac{15}{2} M_i^2 + \frac{25}{4} M_k^2 + 4 \, M_i \, M_k \, A^*_{ik} - 3 \, M_k^2 \, B^*_{ik} \right)}{(M_i + M_k)^2 \; \mathcal{D}_{ik}} \tag{10c}$$

(A^*_{ij}, B^*_{ij} = ratios of reduced collision integrals, see [1]).
Therefore, an empirical "Eucken-Hirschfelder" expression [6,7] is used for the internal part of the mixture heat conductivity (Eq. 9).

If thermal diffusion is disregarded for the moment, the calculation of the multicomponent diffusion fluxes j_i and the multicomponent thermal conductivity λ is now reduced to the determination of binary diffusion coefficients $\mathcal{D}_{ij}$ and of pure species thermal conductivities λ_i. Solutions of systems of n linear equations (n = number of species) are necessary in both cases. Thus, the calculation time is of the order of that for the implicit solution of the instationary conservation equations.

The calculation of the thermal diffusion coefficient $D_{T,i}$ at least demands the solution of systems of 2n linear equations [1]. Since calculation times for this procedure are proportional to $(2n)^3$, the majority of the calculation time would be wasted for the determination of $D_{T,i}$, which seems to be marginal for the problem under consideration. Thus, there is need for a simplified method of calculation of the thermal diffusion coefficient.

2.3 Simplified Transport Models: Diffusion and Heat Conduction

A simplified expression for the diffusion fluxes [8-10] is given by

$$j_i = - D_{i,M}\, \varrho \frac{\partial w_i}{\partial z} \; ; \quad D_{i,M} = \sum_{j \neq i} \frac{1 - w_i}{x_j / \mathcal{D}_{ij}} \qquad (11)$$

This formula can be derived from Eq.(6) for traces of species i. Eq.(11) does not satisfy the balance equation

$$\sum_i j_i = 0 \qquad (12)$$

Therefore, j_i must be derived from Eq.(12) for the respective excess component.

An analogous simplification for the mixture thermal conductivity [11,12] is given by

$$\lambda_M = \frac{1}{2} \left(\sum_i x_i \lambda_i + \frac{1}{\sum_i x_i / \lambda_i} \right) \qquad (13)$$

Fig. 1 gives a comparison of this simple transport model Eq. (11)-(13) with the multicomponent transport model Eq.(6)-(10) for ozone decomposition flames (curves E and M).

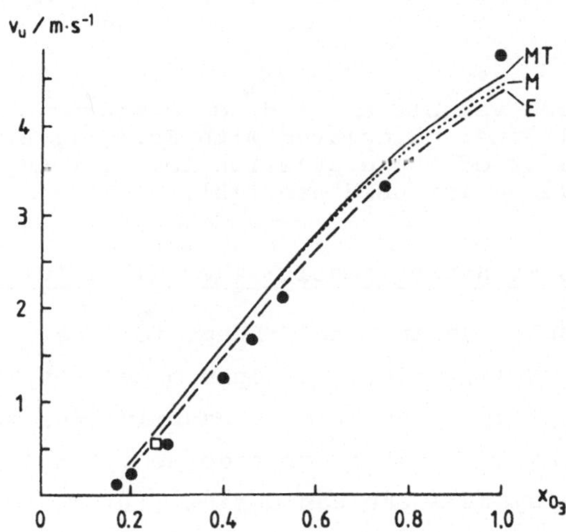

Fig.1: Calculated(lines) and measured(points)flame velocities in O_2-O_3 mixtures; P = 1 bar, T_u=298 K. E: simplified model Eq.(11)-(13),M: multicomponent model Eq.(6)-(10),$D_{T,i}$= 0,MT: like M, thermal diffusion included.Further information in [2].

The detailed form of Eq.(11) plays an important part in the calculation of flame structure, as shown in Fig. 2. Omission of the mass fraction W_i in the numerator of the expression for $D_{i,M}$ in Eq.(11) leads to a completely different form of the concentration dependency of the flame velocity in H_2-O_2 mixtures (the change of the absolute values of v_u should not be overrated).

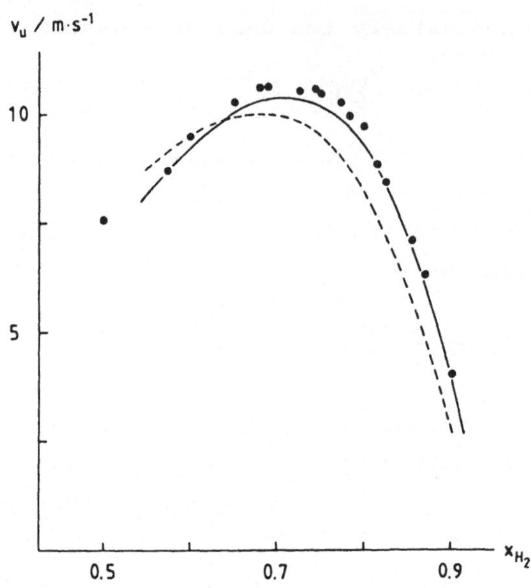

Fig. 2: Free flame velocities in H_2-O_2 mixtures, P = 1 bar, T_u = 298 K. Full line: calculated with Eq.(11); broken line: W_i in the numerator of the expression for $D_{i,M}$ neglected. For details of the mechanism used see [13].

2.4 Simplified Transport Models: Thermal Diffusion

Thermal diffusion is unimportant except for the light species He, H_2 and H atoms [1,14-16]. A simple model for thermal diffusion of species i can be derived by assuming the multicomponent mixture to be a binary mixture of species i and the remainder [14]. For binary systems one can derive [1]

$$D_{T,i} = k_{T,i} \frac{c M_1 M_2}{\rho} \mathcal{D}_{12} \tag{14}$$

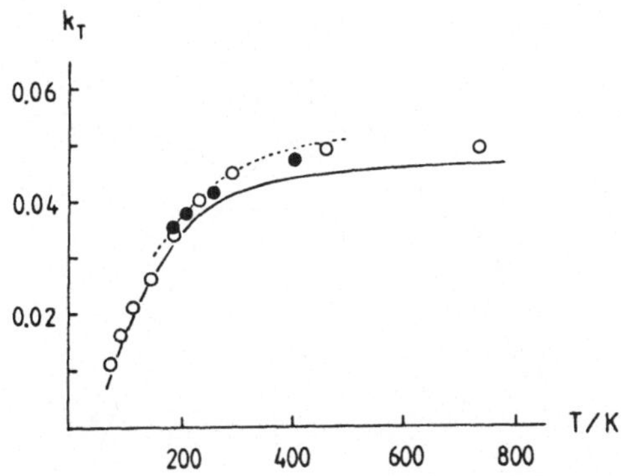

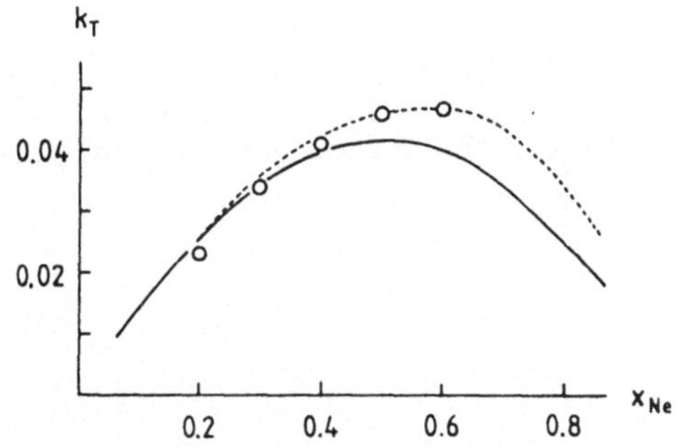

Fig. 3: Thermal diffusion factor k_T as function of temperature and mixture composition in the system Ar-Ne. Full line: calculated with Eq.(15); broken line: calculated with the Chapman-Enskog theory [1]. The points are measurements of different workers (see [1]).

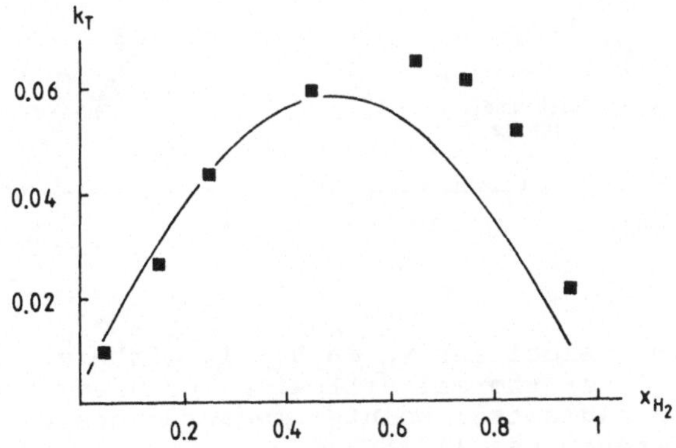

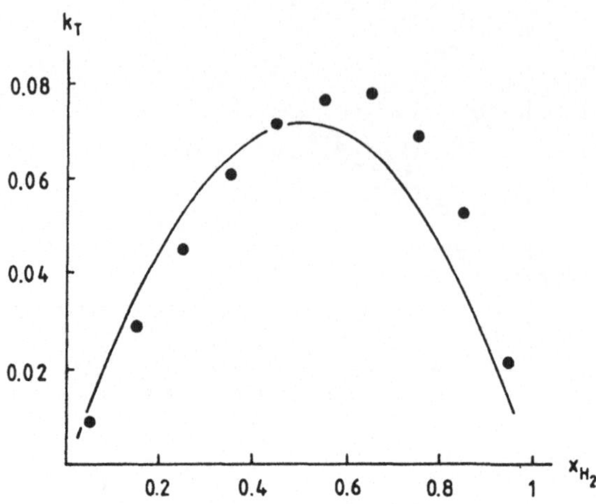

Fig. 4: Thermal diffusion factor k_T as function of the mixture composition ■ in the system H_2-N_2 at 293 K, ● in the system H_2-Ar at 293 K (see [1]). The lines are calculated with Eq.(15).

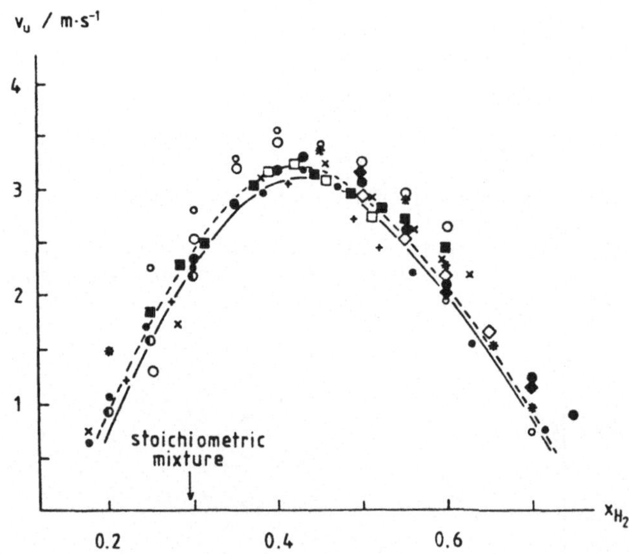

Fig. 5: Free flame velocities v_u in H_2-air mixtures; P = 1 bar, T_u = 298 K. Full line: thermal diffusion included; broken line: thermal diffusion neglected; points: measurements by different workers (for reference see [13]).

Comparison with experiments (see Fig. 3 and 4) shows the validity of the expression [17]

$$k_{T,1} = - k_{T,2} = x_1 x_2 \frac{1}{2} \frac{M_1 - M_2}{M_1 + M_2} k_T^* \qquad (15)$$

with k_T^* being a universal function of the reduced temperature [1]. Apart from the factor of 1/2, Eq.(15) is similar to an expression valid for thermal diffusion of heavy isotopes [1].

The influence of thermal diffusion in ozone decomposition flames is shown in Fig. 2 (difference between curves M and MT). Another example is given in Fig. 5, showing the free flame velocity in H_2-air mixtures at atmospheric pressure.

2.5 Binary Diffusion Coefficients and Thermal Conductivities of Pure Gases

The determination of diffusion fluxes j_i and thermal conductivity λ in multicomponent mixtures is influenced not only be the choice of the transport model used, but also by the specific way of calculation of binary diffusion coefficients $\mathcal{D}_{ij}$ and pure species thermal conductivities λ_i.

$$[\mathcal{D}_{12}]_1 = \frac{3}{16} \frac{\sqrt{2\pi k^3 T^3/\mu}}{P\pi\sigma_{12}^2 \Omega^{(1,1)*}} \qquad (16)$$

$$[\eta]_1 = \frac{5}{16} \frac{\sqrt{\pi m k T}}{\pi\sigma^2\Omega^{(2,2)*}} \qquad (17)$$

$$[\lambda]_1 = \frac{\eta}{M} \left(f_{trans} \bar{C}_{V,trans} + f_{rot} \bar{C}_{V,rot} + f_{vib} \bar{C}_{V,vib} \right) \qquad (18)$$

$$f_{trans} = \frac{5}{2}\left[1 - \frac{2}{\pi} \frac{C_{V,rot}}{C_{V,trans}} \frac{A}{B}\right] \qquad (18a)$$

$$f_{rot} = \frac{\rho D_{rot}}{\eta}\left[1 + \frac{2}{\pi}\frac{A}{B}\right] \qquad (18b)$$

$$f_{vib} = \frac{\rho D_{vib}}{\eta} \qquad (18c)$$

$$A = \frac{5}{2} - \frac{\rho D_{rot}}{\eta} \qquad (18d)$$

$$B = Z_{rot} + \frac{2}{\pi}\left[\frac{5}{3}\frac{\bar{C}_{V,rot}}{R} + \frac{\rho D_{rot}}{\eta}\right] \qquad (18e)$$

Theories for the extrapolation of binary diffusion coefficients (Eq.16) and thermal conductivities of pure monatomic gases (Eq. 18 with $f_{trans} = 5/2$, $f_{rot} = f_{vib} = 0$) are well developed [1]. Eq.(16) to (18) are derived assuming Stockmayer intermolecular potentials:

$$\phi(r) = 4\varepsilon\left[\left(\frac{\sigma}{r}\right)^{12} - \left(\frac{\sigma}{r}\right)^{6}\right] - \frac{2\mu^2}{r^3} \qquad (19)$$

Potential parameters are given elsewhere [2-4, 17,18].

However, there are difficulties with the determination of thermal conductivities of polyatomic species at high temperature due to the contributions of internal degrees of freedom (Eucken correction). For demonstration an extended formulation [17,19] shall be considered (Eq.18).

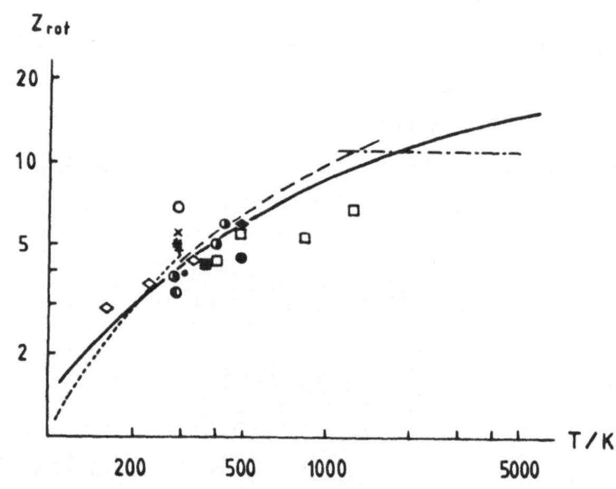

Fig. 6: Rotational collision number of N_2 as function of the temperature. Full line: calculated with the theory of Parker[21], broken line and points: measurements by different workers (for reference see [17]).

This formulation separates the influence of translational,rotational and vibrational degrees of freedom due to their different characteristic temperatures. A little simpler is an expression given by Mason [22], combining rotational and vibrational part to an "internal" part with f_{int} equal to f_{rot}

given in Eq.(18b):

$$[\lambda]_1 = \frac{\eta}{M} (f_{trans} \bar{c}_{V,trans} + f_{int} \bar{c}_{V,int}) \tag{20}$$

The next level of simplification is the assumption of an infinite rotational collision number; $Z_{rot} \rightarrow \infty$. Now Eq.(18) and (20) reduce to the "modified Eucken correction" [1,23]:

$$[\lambda]_1 = \frac{\eta}{M} (\frac{5}{2} \bar{c}_{V,trans} + \frac{\varrho D}{\eta} \bar{c}_{V,int}) \tag{21}$$

Furthermore, the Schmidt number Sc = $\eta / \varrho \cdot D$ can be assumed to be Sc = 1. This level of simplification corresponds to the well known Eucken correction [24].

$$[\lambda]_1 = \frac{\eta}{M} (\frac{5}{2} \bar{c}_{V,trans} + \bar{c}_{V,int}) \tag{22}$$

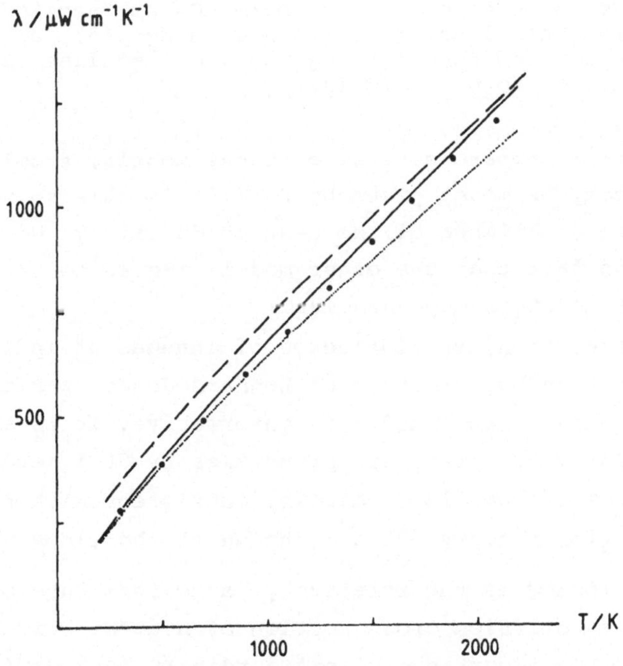

Fig. 7: Thermal conductivity of pure N_2 as function of the temperature. Full line: calculated with Eq.(18), broken line: model of Mason and Monchick, Eq.(20), dashed line: calculated with the Eucken formula, Eq.(22).

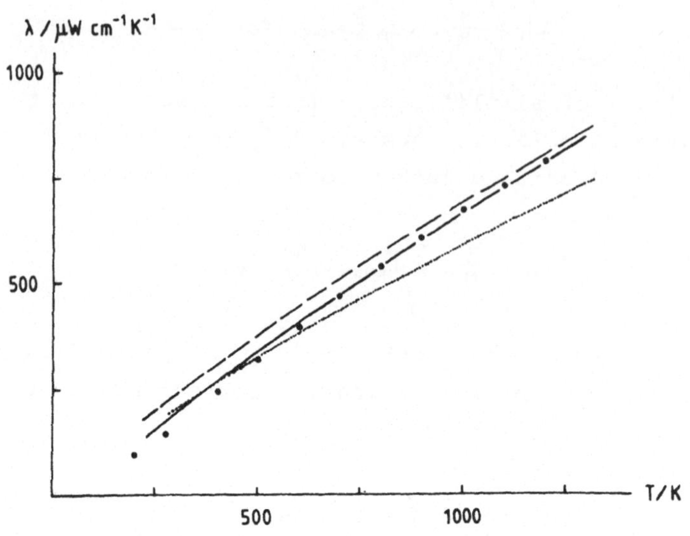

Fig. 8: Thermal conductivity of pure CO_2 as function of the temperature. Full line: calculated with Eq.(18), broken line: model of Mason and Monchick, Eq.(20), dashed line: calculated with the Eucken formula, Eq.(22).

Figs. 7 and 8 compare several of these models, showing that only the most complex model given by Eq.(18) is able to describe the temperature dependence of the heat conductivity. Most important is here the fact that the other models (Eq.20 to 22) give incorrect temperature dependencies.

In this connection, the influence of changes of individual transport properties by variation of Lennard-Jones parameters demonstrated in Tables 1 and 2 is informative. It is shown that a variation of the transport properties by 50 % leads to relative changes of the flame velocity consistent with the result of simple flame theory (22.5 % change of the flame velocity).

Worth mentioning is the relatively large influence of a variation of the collision cross section of H atoms which can be explained by the importance of extraordinary fast diffusion of H atoms into the unburnt gas. The same effect can be noticed in H_2-O_2-N_2 flames.

Variation	New v_u [cm/s]	Deviation [%]
$\delta_O^2 \times 1.5$	137.3	-2.5
$\delta_{O_2}^2 \times 1.5$	115.4	-18.0
$\delta_{O_3}^2 \times 1.5$	143.6	+2.0

Table 1: Influence of variations of Lennard-Jones collision cross sections on the flame velocity of an ozone decomposition flame (40 % O_3, 60 % O_2, P=1 bar, T_u= 298 K, v_u= 140.8 cm/s)

Variation	New v_u [cm/s]	Deviation [%]
$\delta_H^2 / 1.5$	376	+7.5
$\delta_F^2 / 1.5$	351	+0.3
$\delta_{HF}^2 \times 1.5$	349	-0.3
$\delta_{F_2}^2 / 1.5$	350	0
$\delta_{Ar}^2 / 1.5$	396	+13.8
$\delta_{H_2}^2 / 1.5$	351	+0.3

Table 2: Influence of variations of Lennard-Jones collision cross sections on the flame velocity of a H_2-F_2 flame (85 % Ar, 7.55 % H_2, and 7.45 % F_2, P= 1 bar, T_u= 298 K, v_u= 350 cm/s)

3. Influence of Boundary Conditions on Flame Structure

The proper choice of boundary conditions is no substantial problem in the calculation of freely propagating flames (see below), but becomes serious in the calculation of burner-stabilized flames. The study of this process of stabilization of a laminar flat flame on a porous plug burner has attracted much interest since the introduction of this burner type in 1954 [25]. There are mainly four reasons for this interest:

(1) The problem of explaining stabilization on a porous plug burner has inspired a lot of theoretical work, but up to the present no quantitative explanation could be given. Work in the literature [26-30] is confined to qualitative aspects of flame stabilization assuming one-step reaction

kinetics, which cannot describe real flame behaviour. Probably, this deficiency is the reason for a long discussion of the question of the existence of two flame speeds for fixed reactants, cold-boundary state, and heat transfer from the flow at the flame holder. This discussion is supported by conflicting results of corresponding experiments [26,28,31-33].

(2) The problem of flame stabilization is relevant for experimental reasons, since several times flame velocities have been measured with the aid of stabilized flat flames [25,34-36]. Extrapolation from these measurements to zero heat flux to the burner can yield the free flame velocity, but there is the question how to do this extrapolation in a proper way.

(3) The stabilization of a laminar flat flame front is relevant for the decription of the quenching process at a cold wall (for instance, in internal combustion engines in connection with the problem of emission of pollutants).Very interesting is here the development of proper boundary conditions including chemical interaction with the wall.

(4) Last not least, it should be remembered that nearly all work on detailed flame structure is done by use of burner-stabilized flat flames. For proper evaluation of these experiments a quantitative description of the stabilization process is necessary.

The purpose of this chapter is to quantify these statements regarding the role of a porous-plug burner in stabilizing a premixed laminar flame. In particular, the heat transfer to the flame-holder, and information on heterogeneous chemical reactions on the burner surface are sought.

3.1 Boundary Conditions in Burner-Stabilized Flames, Results for a Low Pressure H_2-O_2 Flame

The boundary conditions for the species mass fractions w_i at the surface of a cooled porous plug burner can be derived by integration of the species mass conservation equation (see Eq. 1 to 3 in Section 1.2) between the unburnt gas at $z = -\infty$

(denoted by the index u,see Fig. 9) and the cold boundary of the grid point system (denoted by the index c).

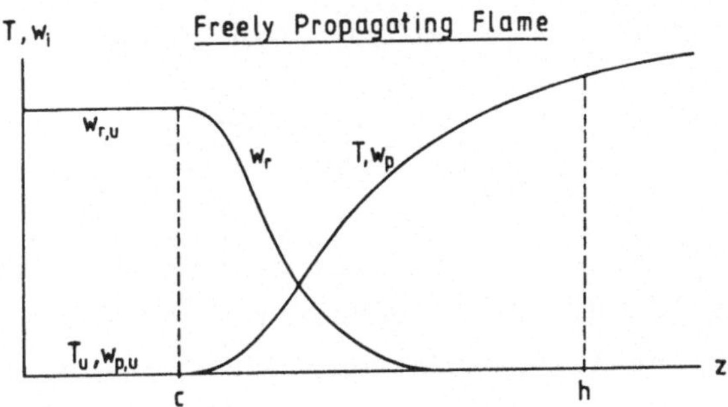

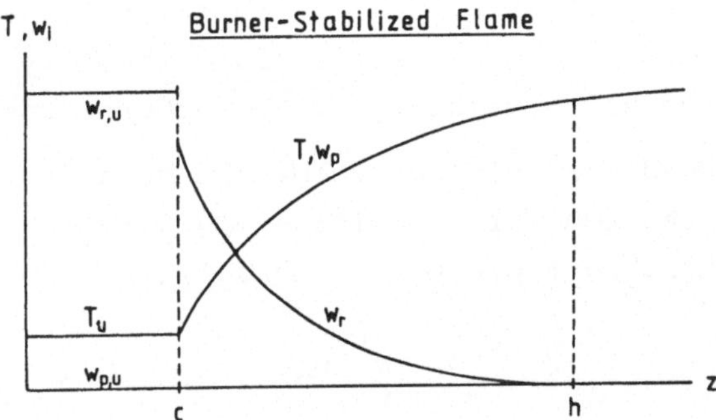

Fig. 9: Illustration of the boundary conditions in freely propagating (upper drawing) and in burner-stabilized flames (see text). Indices: r = reactant , p = product, u = unburnt. At the hot boundary, constant slope of the profiles is assumed[3,13].

Assuming negligible chemical reaction in the cold gas near the cooled burner, and assuming stationarity, the result is the condition of constant mass flux fraction:

$$w_{i,c} = w_{i,u} \quad , \quad \left(\frac{dw}{dz}\right)_h = \text{const} \tag{23}$$

In a freely propagating (adiabatic) flame, the corresponding result of the integration of the enthalpy conservation equation assuming constant c_p near the cold boundary is

$$T_c = T_u - \frac{q}{\rho v \cdot c_{p,c}} \qquad (24)$$

where q_c is the heat flux at the cold boundary c:

$$q_c = \left(-\lambda \frac{dT}{dz}\right)_c + \left(\sum_i j_i h_i\right)_c \qquad (25)$$

In a stationary burner-stabilized flame (see Fig. 9), due to the heat flux to the cooled burner the enthalpy conservation does not hold any longer. Instead, the temperature of a porous plug burner preventing diffusion at the cold boundary c has a fixed value which is maintained by removing the heat flux q_c (see Eq. 25) from the flame by cooling. Stationarity is achieved by the fact that this heat flux q_c into the burner lowers the flame temperature and thereby the flame velocity until its value has reached the given velocity of the unburnt gas.

If it is assumed that cooling keeps the burner at the cold gas temperature, the boundary condition (Eq.24) for freely propagating flames must be replaced by the relation

$$T_c = T_u \qquad (26)$$

Application of these boundary conditions (Eq. 23 and 26) to a burner-stabilized rich H_2-O_2 low-pressure flame (see Fig. 10) predicts an H atom mole fraction of 15 percent in the mixture at the burner surface. In contray, experiments by Wagner and coworkers [37] show the H atom mole fraction near the burner surface to be near zero.

The most simple and reasonable explanation of this discrepancy is the occurence of instantaneous recombination of the quickly diffusing H atoms at the burner surface. This leads to the following modification of the boundary conditions:

$$w_{i,c} = w_{i,u} - \frac{j_{i,c}}{\rho v} \qquad \text{(except for } H, H_2\text{)} \qquad (27a)$$

$$w_{H,c} = w_{H,u} = 0 \qquad \text{(for H atoms)} \tag{27b}$$

$$w_{H_2,c} = w_{H_2,u} - \frac{j_{H_2,c}}{\rho v} - \frac{j_{H,c}}{\rho v} \qquad \text{(for } H_2\text{)} \tag{27c}$$

The mass fraction $w_{H,c}$ of the H atoms at the cold boundary (say: burner surface) is set to zero. The recombined H atoms are then taken into account by an additional term in the boundary condition for the recombination product H_2.

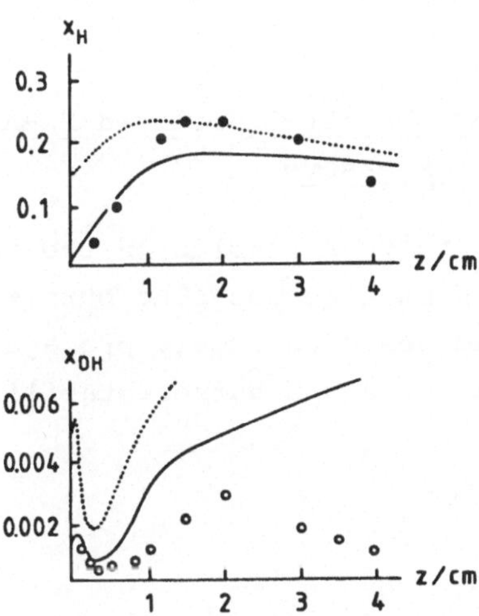

Fig. 10: Measured and calculated mole fraction profiles of H and OH in a burner-stabilized rich H_2-O_2 flame; P = 13.9 mbar, T_u = 298 K, $x(H_2)$ = 0.75, $x(O_2)$ = 0.25, v_u = 178 cm/s. Points: experimental and results [37], dotted lines: calculated without heteregeneous recombination, full lines: calculated assuming heterogeneous recombination at the burner surface.

Now there is agreement of calculated and measured H atom profiles within the limits of experimental errors (see Fig. 10). Correspondingly, for the OH radical there is agreement, too,in the near of the burner; discrepancies at large distances from the burner surface are caused by cooling of the experimental flame by the surrounding cold gas.

An important and instructive quantity is the heat flux $\Delta q_{c,u}$ to the burner given by

$$\Delta q_{c,u} = \left(-\lambda \frac{dT}{dz}\right)_c + \left(\sum_i j_i h_i\right)_c + \rho v\left[\left(\sum_i w_i h_i\right)_c - \left(\sum_i w_i h_i\right)_u\right] \qquad (28)$$

since it can be compared easily with experimental results.

For the rich H_2-O_2 low-pressure flame illustrated in Fig. 10, it can be calculated that only 21 percent of the heat flux $\Delta q_{c,u}$ to the burner are due to thermal conduction, whereas the rest of 79 percent is caused by the heat of heterogeneous recombination of H atoms !

3.2 Results for Burner-Stabilized H_2-Air Flames at Atmospheric Pressure

In this section some further results on the stabilization process of H_2-air flames on a porous plug burner shall be demonstrated. Typical examples of a freely propagating (this is the flame of test problem b) and a burner-stabilized flame are given in Fig. 11.

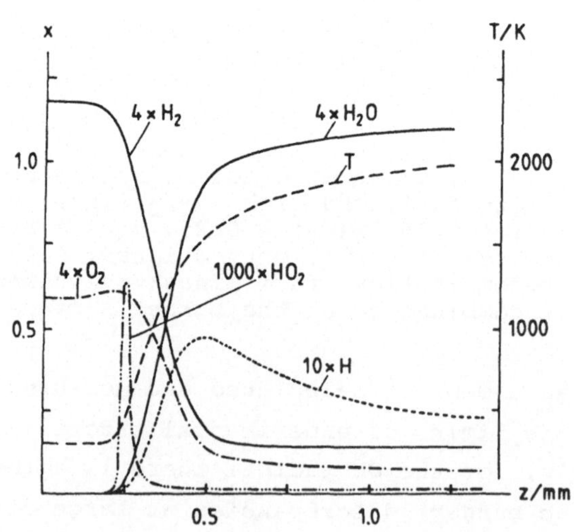

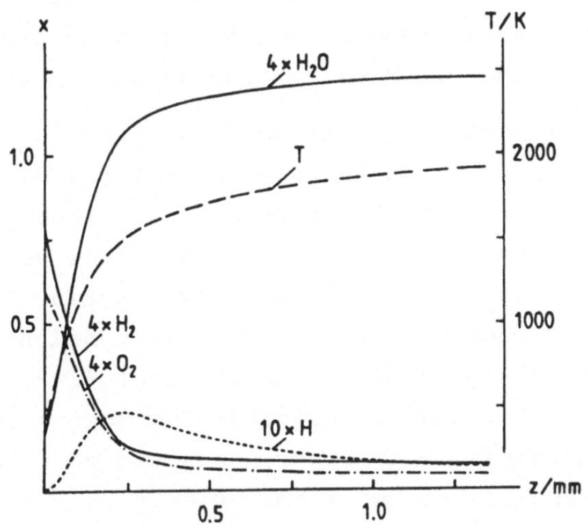

Fig. 11: Mole fraction and temperature profiles in a freely propagating (upper illustration) and in a burner-stabilized (lower illustration) H_2-air flame. P = 1 bar, T_u= 298 K, stoichiometric mixture; stabilization at v_u= 120 cm/s.

A result of a systematic variation of the unburnt gas velocity v_u is given in Fig. 12.

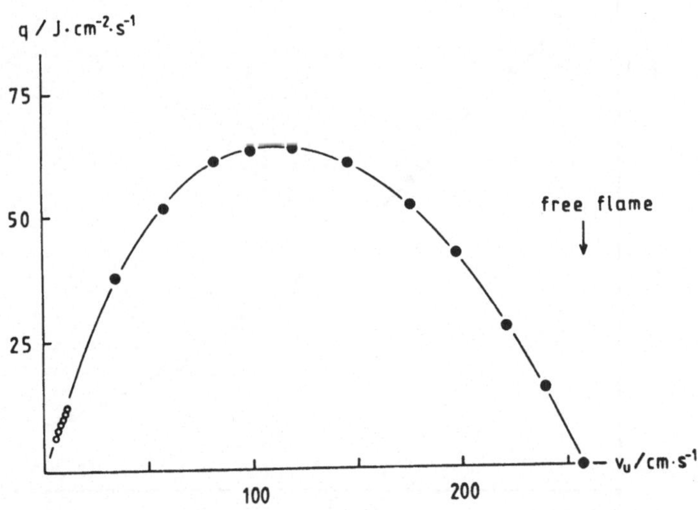

Fig. 12: Heat flux q (per unit time and unit area)as function of the unburnt gas flow velocity v_u for a stoichiometric H_2-air flame at P = 1 bar. -●—●— calculations, o measurements [25].

This illustration shows the heat flux q to the burner per unit time and unit area as function of the unburnt gas flow velocity v_u for a stoichiometric H_2-air flame at atmospheric pressure. At the critical mass flow rate corresponding to the free flame velocity 260 cm/s (calculated negclecting thermal diffusion) there is zero heat flux to the burner. The heat flux q then is increased by decrease of the unburnt gas flow velocity, but then after passing a maximum going down, again. The small points represent experimental results by Spalding and Botha [25] in this interesting region of decreasing heat flux, which is caused by a decreasing temperature gradient due to the increase of the flame front thickness.Both experiments and calculations demonstrate the existence of two flame velocities for a distinct range of heat flux values q.

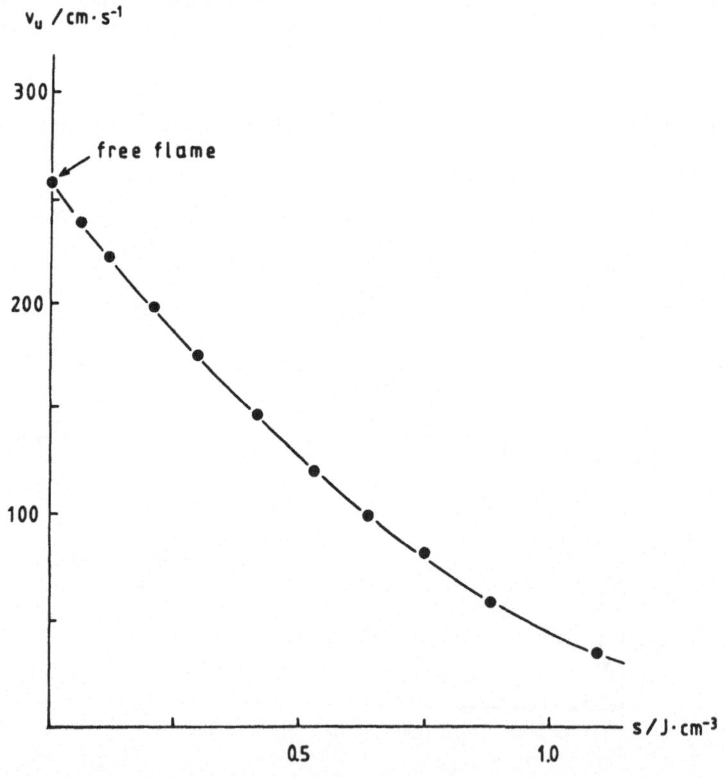

Fig. 13: Calculated unburnt gas flow velocity v_u as function of the heat flux s(per unit time and unit volume of fuel) for the flame described in Fig. 12.

Fig. 13 shows another way of plotting these results, which is usual for the determination of free flame velocities by extrapolation to zero heat flux [25,34-36]. It shows the unburnt gas flow velocity v_u as function of the heat flux s per unit time and unit volume of fuel. This figure illustrates that linear extrapolation leads to small values of the free flame velocity. This extrapolation procedure will lead to correct results only by inclusion of very small heat fluxes s, whose experimental realization is rather difficult.

Fig. 14 demonstrates the corresponding monotonical decrease of the final temperature and the maximum mole fractions of H and O atoms and of OH radicals connected with the reduction of the unburnt gas flow velocity.

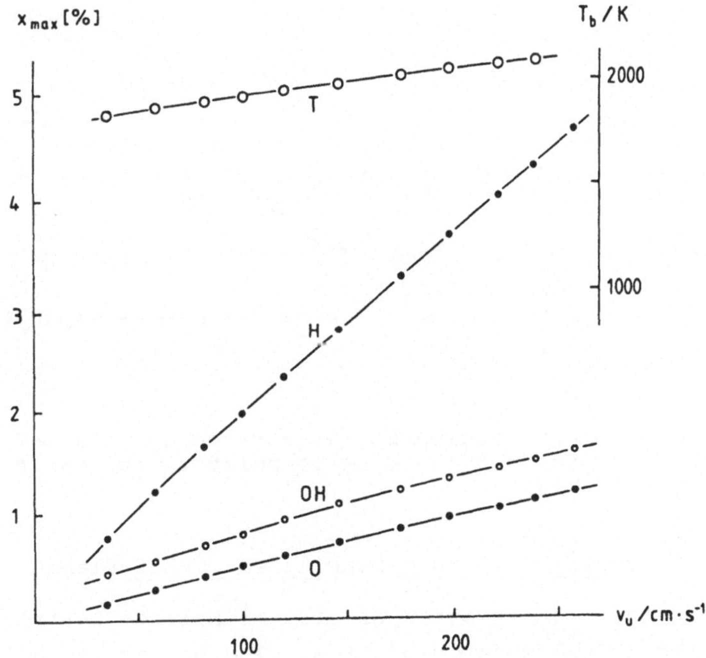

Fig.14: Final temperature (at z = 0.3 cm) and maximum mole fractions of H, O,and OH as function of the unburnt gas flow velocity in the flame described in Fig. 12.

Furthermore,(as demonstrated in Fig. 15), reduction of the unburnt gas flow velocity at first leads to a decrease of the distance z between burner surface and flame front (defined by the position of maximum mole fraction of H atoms). This distance again increases after passing a minimum value corresponding to maximum heat flux to the flameholder (see Fig. 12).

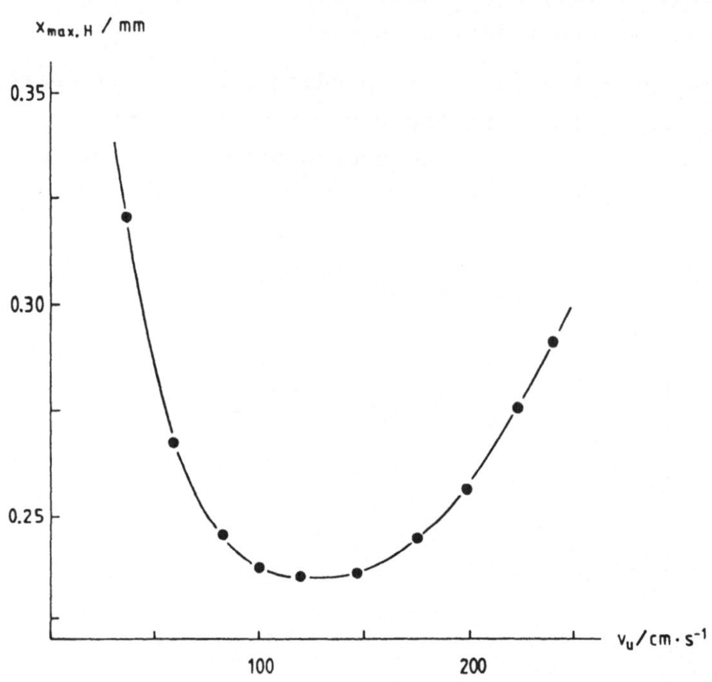

Fig. 15: Distance z between burner surface and flame front as function of the unburnt gas flow velocity v_u for the flame described in Fig. 12.

3.3 Results for Burner-Stabilized CO-Air Flames

Results on the stabilization of stoichiometric CO-air flames at atmospheric pressure on a porous plug burner are given in Fig. 16. The results correspond to that shown in Fig. 12 for H_2-air flames. This time, comparable measurements [36] are existing in the region of maximum heat flux q to the burner, and the agreement of measured and calculated heat fluxes is satisfactory, if the limits of experimental errors are taken into consideration.

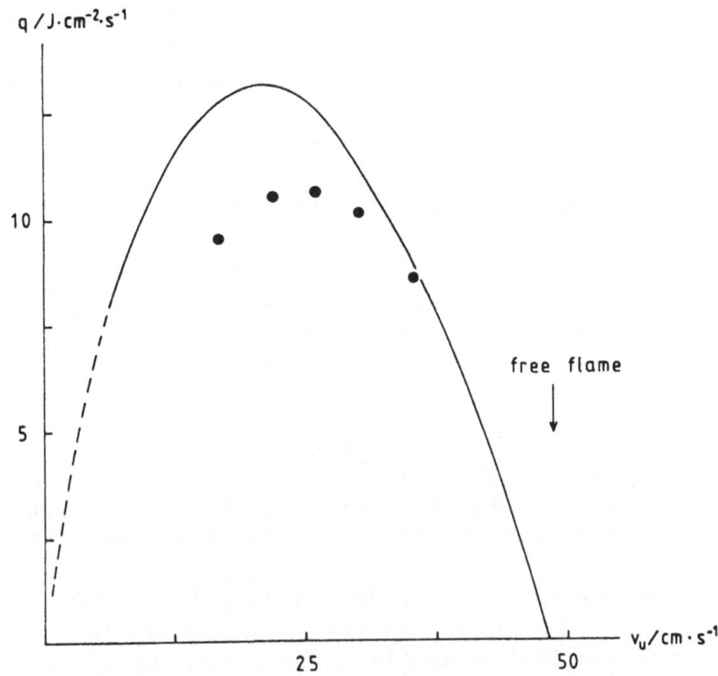

Fig. 16: Heat flux q (per unit time and unit area) as function of the unburnt gas flow velocity v_u for a stoichiometric CO-air flame (8 % H_2 added) at P = 1 bar. Points: measurements [36], line: present calculations.

Acknowledgement

The author thanks Prof. Dr. K.H. Homann and Prof. Dr. H.Gg. Wagner for their sustained interest in this work and many helpful discussions. The financial support of the "Deutsche Forschungsgemeinschaft" and the "Fonds der Chemischen Industrie" is gratefully acknowledged.

References

[1] J.O. Hirschfelder, C.F. Curtiss, R.B. Bird, Molecular Theory of Gases and Liquids. Wiley, New York (1954)

[2] J. Warnatz, Ber.Bunsenges.Phys.Chem. 82, 193 (1978)

[3] J. Warnatz, Ber.Bunsenges.Phys.Chem. 83, 950 (1979)

[4] J. Warnatz, Eighteenth Symposium (International) on Combustion, p. 369. The Combustion Institute, Pittsburgh(1981)

[5] G. Dixon-Lewis, Proc.Roy.Soc. A307, 111 (1968)

[6] J.O. Hirschfelder, Sixth Symposium (International) on Combustion, p. 351. Reinhold, New York (1957)

[7] V. Parkinson, P. Gray, J.Chem.Soc.Faraday I 68,1065,1077, (1972)

[8] J. Stefan, Sitzungsberichte Akad.Wiss.Wien II 68, 325(1874)

[9] J.O. Hirschfelder, C.F. Curtiss, Third Symposium on Combustion, Flame and Explosion Phenomena, p.121. Williams and Wilkins, Baltimore (1949)

[10] H. Watts, Trans.Faraday Soc. 60, 1745 (1964)

[11] J.H. Burgoyne, F. Weinberg, Fourth Symposium on Combustion, p.294. Williams and Wilkins, Baltimore (1953)

[12] S. Mathur, P.K. Tondon, S.C. Saxana, Mol.Phys.12, 569(1967)

[13] J. Warnatz, Ber.Bunsenges.Phys.Chem. 82, 643 (1978)

[14] R.M. Fristrom, A.A. Westenberg, Flame Structure. McGraw-Hill, New York (1965)

[15] D.A. Frank-Kamenetskii, Diffusion and Heat Exchange in Chemical Kinetics. Princeton University Press, Princeton (1955)

[16] G. Dixon-Lewis, Proc.Roy.Soc. A 307, 111 (1968)

[17] J. Warnatz, Berechnung der Flammengeschwindigkeit und der Struktur von laminaren flachen Flammen. Habilitationsschrift, Technische Hochschule Darmstadt (1977)

[18] J. Warnatz, Ber.Bunsenges.Phys.Chem. 82, 834 (1978)

[19] W.F. Ahtye, J.Chem.Phys. 57, 5542 (1972)

[20] C.S. Wang Chang, G.E. Uhlenbeck, J. de Boer, Studies in Statistical Mechanics II, Wiley, New York (1964)

[21] J.G. Parker, Phys.Fluids 2, 449 (1959)

[22] E.A. Mason, L. Monchick, J.Chem.Phys. 36, 1622 (1962)

[23] K. Schäfer, Z.physik.Chem. B 53, 149 (1943)

[24] A. Eucken, Physik Z. 14, 324 (1913)

[25] J.P. Botha, D.B. Spalding, Proc.Roy.Soc. A 225,71(1954)

[26] D.B. Spalding, Proc.Roy.Soc. A 240, 83 (1957)

[27] G.F. Carrier, F.E. Fendell, W.B. Bush, Comb.Sci.Technol. 18, 33 (1978)

[28] C.R. Ferguson, J.C. Keck, Comb.Flame 34, 85 (1979)

[29] A.C. McIntosh, J.F. Clark, Proc.Roy.Soc. A 372.367(1979)

[30] S. Galant, Eighteenth Symposium (International) on Combustion, p. 1343. The Combustion Institute, Pittsburgh (1981)

[31] D.B. Spalding, V.S. Yumlu ,Comb.Flame 4, 553 (1960)

[32] W.E. Kaskan, Comb.Flame 4, 285 (1960)

[33] P.H. Kydd, W.I. Foss, Comb.Flame 8, 267 (1964)

[34] J.L.J. Rosenfeld, T.M. Sudgen, Comb.Flame 8, 37 (1964)

[35] E. Edmondson, M.P. Heap, R. Pritchard, Comb.Flame 14, 195 (1970)

[36] V.S. Yumlu , Comb.Flame 11, 389 (1967)

[37] K.H. Eberius, K. Hoyermann, H.Gg. Wagner, Thirteenth Symposium (International) on Combustion, p. 713. The Combustion Institute, Pittsburgh (1971)

NUMERICAL SOLUTION OF BURNER-STABILIZED PRE-MIXED LAMINAR FLAMES BY AN EFFICIENT BOUNDARY VALUE METHOD*

Mitchell D. Smooke
Applied Mathematics Division

James A. Miller
Combustion Physics Division

Robert J. Kee
Applied Mathematics Division

SANDIA NATIONAL LABORATORY
Livermore, CA USA

ABSTRACT

A numerical technique has been developed for integrating the one-dimensional, steady-state, pre-mixed laminar flame equations. A global finite difference approach is used in which the nonlinear difference equations are solved by a damped-modified Newton method. An assumed temperature profile helps to generate a converged numerical solution on an initial coarse grid. Mesh points are inserted in regions where the solution profiles exhibit high gradient and high curvature activity. These features are discussed and illustrated in the paper.

1. Introduction

Laminar pre-mixed flames (flat flames) are commonly used to investigate chemical kinetics processes that are important in combustion. Because of its essentially one-dimensional nature, the flat flame is particularly useful in constructing computational models. These models can be used in close conjunction with experimental data to provide detailed information on flame structure and elementary reaction paths.

The calculation of laminar pre-mixed flame structure was one of the first combustion problems to be attacked by both analytical and numerical techniques. The problem is posed most concisely as a nonlinear two-point boundary value problem and, while the problem has a very simple flow

---

*Prepared by Sandia National Laboratory, Livermore, CA, 94550 for the United States Department of Energy.

configuration, the direct solution of the governing equations has proven to be difficult and is still being actively pursued.

Although the literature contains numerous examples of the use of time dependent methods to investigate flat flame structure (e.g. [1-4]), there are relatively few references to the solution of the flame equations by steady-state methods [5-7]. In this paper we discuss the application of an efficient boundary value method to the solution of burner-stabilized (nonadiabatic) pre-mixed laminar flames. The method we consider is a variation of a global finite difference method. It enables us to obtain "good" starting estimates for the temperature and species profiles, and it enables us to place grid points accurately in regions of high gradient activity of the dependent solution components.

In the next section we introduce some notation and formulate the pre-mixed flame problem as a nonlinear two-point boundary value problem. Section 3 presents the boundary value method we have used to solve the flame equations and Section 4 contains the results of applying our method to several laboratory flames.

2. Problem Formulation

Our formulation of the burner-stabilized flame problem closely follows that originally proposed by Hirschfelder and Curtiss [8]. The physical problem is illustrated in Figure 1. A pre-mixed fuel and oxidizer mixture flows through a cooled porous plug burner. As the mixture emerges from the burner, it passes through a reaction zone in which chemical changes take place. Further downstream, it eventually emerges in a burned state. Our goal is to be able to predict theoretically Y_k, $k = 1,2,...,N$, the mass fractions of the species under investigation, and T, the temperature of the combustible mixture, as functions of the height x above the burner.

Our formulation of the problem assumes the following:
(i) The flow is one-dimensional and the region under

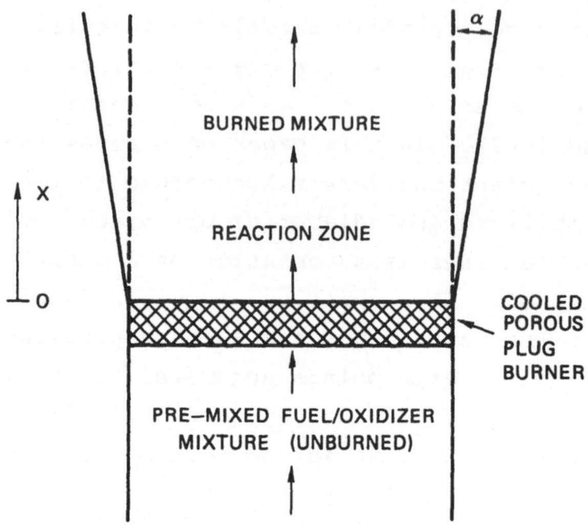

Figure 1. Schematic of nonadiabatic pre-mixed laminar flame. The flame gases expand as a function of the height x above the burner; the angle α is the expansion angle.

consideration is the semi-infinite interval $0 \leq x < \infty$. (ii) The thermodynamic pressure p is constant. (iii) We allow for the possible quasi-one-dimensional expansion of the flame gases, where A(x) represents the cross sectional area of the flame as a function of the height x above the burner. (iv) Body forces are negligible. (v) Radiative heat transfer is negligible. (vi) The diffusion of heat due to concentration gradients (Dufour effect) is negligible.

With approximations (i)-(vi) the equations governing the structure of a steady one-dimensional isobaric flame are

Continuity:

$$\dot{M} = \rho u A = \text{constant} \quad , \tag{2.1}$$

Conservation of Species:

$$\dot{M}\frac{dY_k}{dx} = -\frac{d}{dx}(\rho A\, Y_k V_k) + \dot{w}_k\, A\, W_k, \qquad k = 1,2,\ldots,N, \quad (2.2)$$

Conservation of Energy:

$$\dot{M}\frac{dT}{dx} = \frac{1}{C_p}\frac{d}{dx}\left(\lambda A \frac{dT}{dx}\right) - \frac{A}{C_p}\left[\sum_{k=1}^{N} \rho\, Y_k V_k C_{pk} \frac{dT}{dx}\right] \quad (2.3)$$

$$- \frac{A}{C_p}\sum_{k=1}^{N} \dot{w}_k\, h_k\, W_k\,,$$

Equation of State:

$$\rho = \frac{\overline{W}p}{RT}\,, \quad (2.4)$$

with the boundary conditions

$$T(0) = T_b, \quad (2.5)$$

$$Y_k + \frac{\rho Y_k V_k A}{\dot{M}} = \varepsilon_k, \qquad k = 1,2,\ldots,N, \quad (2.6)$$

$$\frac{dT}{dx} = 0, \qquad \text{as } x \to \infty, \quad (2.7)$$

$$\frac{dY_k}{dx} = 0, \qquad \text{as } x \to \infty, \qquad k = 1,2,\ldots,N. \quad (2.8)$$

In addition to the variables already defined, u denotes the velocity of the fluid mixture, ρ the density of the fluid mixture, W_k the molecular weight of the kth species, $\overline{W}$ the mean molecular weight of the mixture, R the universal gas constant, λ the thermal conductivity of the mixture, C_p the specific heat capacity of the mixture at constant pressure, C_{pk} the specific heat capacity at constant pressure of the kth species, $\dot{w}_k$ the rate of production of the kth species by chemical reaction, h_k the specific enthalpy of the kth species with respect to the mixture of remaining species, ε_k the known incoming mass flux fractions, and V_k the binary diffusion velocity of the kth species.

The diffusion velocity is divided into three parts

$$V_k = v_k + w_k + v_c, \qquad k = 1,2,\ldots,N, \quad (2.9)$$

where v_k is the ordinary diffusion velocity due to mole fraction gradients, w_k is the thermal diffusion velocity and v_c is a constant diffusion velocity (independent of species) which is added to insure that mass is conserved. We approximate v_k by the Curtiss-Hirschfelder [9] approximation. We write

$$v_k = -\frac{1}{X_k} D_k \nabla X_k \,, \quad k = 1,2,\ldots,N, \tag{2.10}$$

where X_k is the mole fraction of the kth species and D_k is related to the binary diffusion coefficients $\mathscr{D}_{jk}$ through the relation

$$D_k = \frac{1 - Y_k}{\sum_{j \neq k} X_j / \mathscr{D}_{jk}} \,. \tag{2.11}$$

Thermal diffusion is incorporated into the model only in the trace light component limit. The thermal diffusion velocity is given by

$$w_k = \frac{D_k \, t_{D_k}}{X_k} \frac{1}{T} \frac{dT}{dx} \,, \tag{2.12}$$

where t_{D_k} is the thermal diffusion coefficient of species k. We only apply Eq. (2.12) for low molecular weight species such as H, H_2 or He. The constant diffusion velocity v_c is introduced in order to satisfy the condition

$$\sum_{k=1}^{N} V_k Y_k = 0, \tag{2.13}$$

which must be satisfied if mass is guaranteed to be conserved. Upon making use of Eqs. (2.9) and (2.13) we have

$$v_c = -\sum_{k=1}^{N} Y_k \, (w_k + v_k). \tag{2.14}$$

The transport model we use is essentially the same as that recommended by Coffee and Heimerl [10]. The multicomponent conductivity is computed from the single component conductivities using Wilke's semi-empirical formula [11], and an approximate Eucken-Hirschfelder correction for polyatomic species is incorporated in the single component conductivities in the manner followed by Svehla [12]. The thermal diffusion coefficients are calculated using the thermal diffusion ratio discussed in Chapman and Cowling [13].

3. Method of Solution

The pre-mixed flame problem is formulated as a nonlinear two-point boundary value problem on the semi-infinite interval $0 \leq x < \infty$. The singularity at $x = \infty$ is handled by solving the governing equations on the finite interval $0 \leq x \leq L$ for some number $L < \infty$ with the boundary conditions at ∞ imposed at $x = L$. The value of L can be obtained from experiment or from numerical computation. Our finite difference formulation follows by applying appropriate difference expressions to the derivatives in Eqs.(2.1-2.8) on a mesh $\mathcal{M}$, where h_j, $j=1,2,...,M$, denotes the distance between consecutive mesh points. We convert the process of finding an analytic solution to Eqs. (2.1-2.8) to one of finding a discrete solution, Z_h^*, of the resulting nonlinear system of difference equations which we denote by

$$F(Z_h^*) = 0. \tag{3.1}$$

For an initial solution estimate Z_h^0 which is sufficiently close to Z_h^*, the system of nonlinear equations (3.1) can, in principle, be solved by a variety of nonlinear equation methods. A common approach is to apply a version of Newton's method, which we write in the form

$$Z^{n+1} = Z^n - \lambda^n J^{-1}(Z^n) F(Z^n). \tag{3.2}$$

Here Z^i denotes the ith solution iterate, λ^i the ith damping parameter $(0 < \lambda \leq 1)$ and $J(Z^i) = \partial F(Z^i)/\partial Z$ the $(N+1)(M+1) \times (N+1)(M+1)$ Jacobian matrix. The choice of the damping parameter is discussed in detail by Deuflhard [14]. We have implemented a variation of his method for nonsingular Jacobians.

In practice, (3.2) is rewritten in the form

$$J(Z^n)(Z^{n+1}-Z^n) = -\lambda^n F(Z^n), \tag{3.3}$$

where at each iteration a system of linear equations is solved for corrections to the previous solution vector. For flame calculations in which the cost of forming the Jacobian matrix--either analytically or numerically--is a significant part of the calculation, we apply a modified version of Newton's method in which the Jacobian is only re-evaluated periodically.

Although the use of a damped-modified Newton method in the numerical solution of two-point boundary value problems is not unusual, there are, however, several features of our implementation which warrant further discussion.

Assumed Temperature Profile

We note that if the pre-mixed flame problem defined in Eqs. (2.1-2.8) is solved with an assumed temperature profile, the energy equation can be replaced by $T = T(x)$, where $T(x)$ is known. One advantage to such a procedure is that if an experimental profile is available, substituting this profile for the energy equation produces a solution of the species equations that is a better representation of the chemistry actually occurring in the flame than if the species equations were solved coupled with the energy equation. The reason is that in actual laboratory applications there are always distributed heat losses and the temperature predicted by solving the full set of flame equations is often not a good representation of the temperature profile actually obtained in the laboratory (see Section 4). In addition, in cases where one requires a solution to the full pre-mixed flame equations--predicted temperatures and species profiles--the often severe difficulties associated with the convergence of Newton's method can largely be circumvented by first solving the species equations with a specified temperature profile and then solving the full problem with this solution as an initial estimate. The specification of a temperature profile eliminates the exponential nonlinearities in temperature and the resulting chemical production terms contain nonlinearities that are at most algebraic in the species concentrations.

Upwind Differencing

The convective derivatives in Eqs. (2.1-2.8) have all been differenced using upwind difference approximations. Due to the truncation error in the upwind difference approximations for the convective terms in Eqs. (2.1-2.8) we are, in effect (through first order in h), solving the original flame equations with an extra diffusion term, the size of which is proportional to $\rho u h_j/2$. As a result, the numerical solutions can be broadened depending on the size of $\rho u h_j/2$ compared to λ/Cp and ρD_k, $k=1,2,\ldots,N$. Since the coefficient of the numerical diffusion term is proportional to h_j, we expect the broadening to be most significant on the initial coarse grid. The idea of broadening the solution profiles can be looked upon as making the numerical solution more nearly constant. As a result, one expects to have less difficulty in specifying an initial solution estimate for which the Newton iteration converges than if the high gradient behavior of the numerical solution is preserved.

Adaptive Gridding

A variety of procedures have been used in recent years to obtain optimal grid spacings on which to solve two-point boundary value problems (e.g. [15-17]). The approach we have chosen to determine an adaptive grid for the pre-mixed flame problem is similar to the method used by Pearson [17] in solving scalar boundary layer problems. We attempt to subequidistribute the difference in the components of the discrete solution and the difference in the gradient of the components of the discrete solution between adjacent mesh points. Upon denoting the (N+1) vector S such that

$$S^t = [T, Y_1, Y_2, \ldots, Y_N], \tag{3.4}$$

we seek to obtain a mesh $\mathcal{M}$ such that

$$\int_{x_j}^{x_{j+1}} \left|\frac{dS_i}{dx}\right| dx \leq \delta \left(\max_{0 \leq x \leq L} \left|S_i\right| \right), \quad j=0,1,\ldots,M-1, \quad i=1,2,\ldots,N+1, \tag{3.5}$$

and

$$\int_{x_j}^{x_{j+1}} \left|\frac{d^2S_i}{dx^2}\right| dx \leq \gamma \left(\max_{0 \leq x \leq L} \left|\frac{dS_i}{dx}\right| \right), \quad j=1,2,\ldots,M-1, \quad i=1,2,\ldots,N+1, \tag{3.6}$$

where δ and γ are small numbers less than one and the values of max $|S_i|$ and max $|dS_i/dx|$ are obtained from a converged numerical solution on a previously determined mesh.

In our adaptive mesh algorithm we first solve the flame equations on a coarse mesh (3-5 subintervals). If either of the inequalities is not satisfied, a grid point is inserted at the midpoint of the interval in question. Once a new mesh has been obtained, the previously converged numerical solution is interpolated onto this new mesh. This result serves as an initial solution estimate for the flame equations on the new mesh. The procedure is continued until the inequalities in Eqs. (3.5-3.6) are satisfied to some specified tolerance.

4. Numerical Results

This section discusses the results obtained by applying our method to several laboratory flames. Before proceeding, however, it is worthwhile to discuss a few points concerning the numerical implementation of the method.

We use "S" shaped and Gaussian curves for the initial solution profiles (see [18] for details). Collision integrals are approximated by polynomial fits to the tabulated data of Monchick and Mason [19]. Stockmayer potentials are used throughout in evaluating transport properties, and the potential parameters are compiled from various tabulated sources (e.g. [12]).

Thermodynamic properties (heat capacities, entropies, and enthalpies) are computed from fits of the JANNAF data [20] used in the NASA chemical equilibrium code [21]. In our computations all the chemical production rate terms, thermodynamic properties, and equation of state variables are evaluated using CHEMKIN, the chemical kinetics code package written by Kee, Miller, and Jefferson [22].

With the difference approximations we have chosen to employ, Newton's equations can be written in block tri-diagonal form. We solve these equations using DECBT and SOLBT written by Hindmarsh [23]. We terminate the Newton iteration when

$$||\Delta z^{n+1}||_2 = ||z^{n+1} - z^n||_2 \leq TOL, \; n=0,1,2,..., \tag{4.1}$$

where we typically take $TOL \leq 10^{-5}$.

Test Flames

We have applied the boundary value method to a variety of laboratory flames. In the remainder of this section we discuss the results of our calculations for a pre-mixed hydrogen-oxygen, a pre-mixed methane-oxygen and a pre-mixed hydrogen-air flame.

Hydrogen-Oxygen Flame (H_2/O_2)

Eberius, Hoyermann, and Wagner [24] have experimentally studied a rich, low-pressure, hydrogen-oxygen flame. The experiment was performed at 10.6 torr and the incoming flow velocity was 178 cm/sec. The unburned mixture was 75 mole percent H_2 and 25 mole percent O_2. The reaction mechanism contained 17 reactions and there were eight species under investigation.

Figure 2 shows the experimental and calculated (solid line) OH profiles for the hydrogen-oxygen system. We also include the experimental temperature data. The solid line connecting the temperature points was obtained by a least squares fit to the data. The line has been extended past 4-cm such that a zero temperature gradient is obtained as we approach 10-cm. The predicted OH profile agrees well with the data.

In Figure 3, however, we show the solution in which the energy equation is solved coupled with the species equations. The initial temperature profile is just the extended experimental temperature profile linearly interpolated to the initial coarse grid. In this case the calculated OH concentration and the calculated temperature profile (solid lines) are much higher than observed experimentally. This calculation illustrates two points. First, use of an experimental temperature profile instead of a solution to the energy equation can allow physically realistic chemical kinetic modelling of the flame; and second, the solution method presented above allows one to obtain a solution to the energy equations even when the initial guess is bad (compare Figures 2 and 3).

Both calculations were performed using a 10-cm interval with an initial grid consisting of five equi-spaced subintervals. The calculation with the experimental temperature profile required 45 adaptively placed grid points in order to obtain three significant figures of accuracy. The coupled energy-species calculation required 34 points to achieve the same degree of accuracy. For the first calculation, 22 of the 45 points were located in the flame zone --the region where $500 \leq T \leq .9\ T_{max}$. In the second calculation 25 of the 34 grid points were in the flame zone. The calculations were performed without thermal diffusion and the cross sectional area of the flame was taken as constant. The first calculation took 20 seconds of CPU time on a CRAY-I while the second took 33.

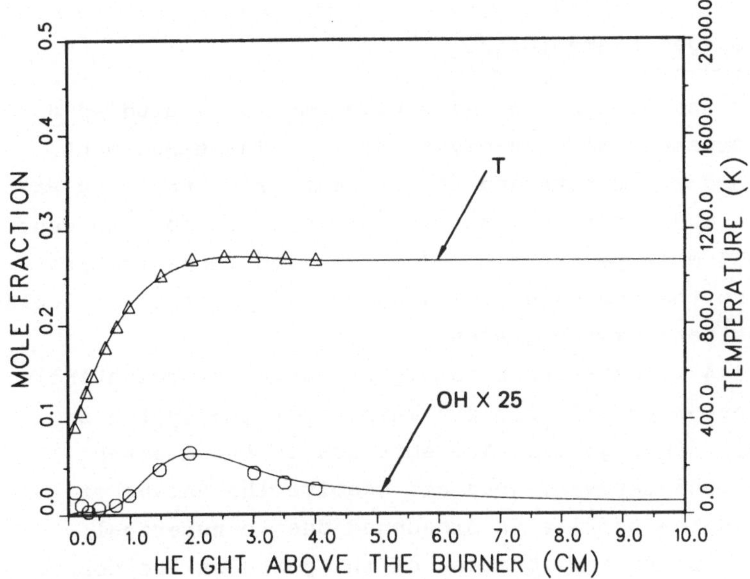

Figure 2. Experimental and calculated (solid line) OH profiles for an H_2/O_2 flame. The calculation was performed by replacing the energy equation with the extended experimental temperature profile (solid line).

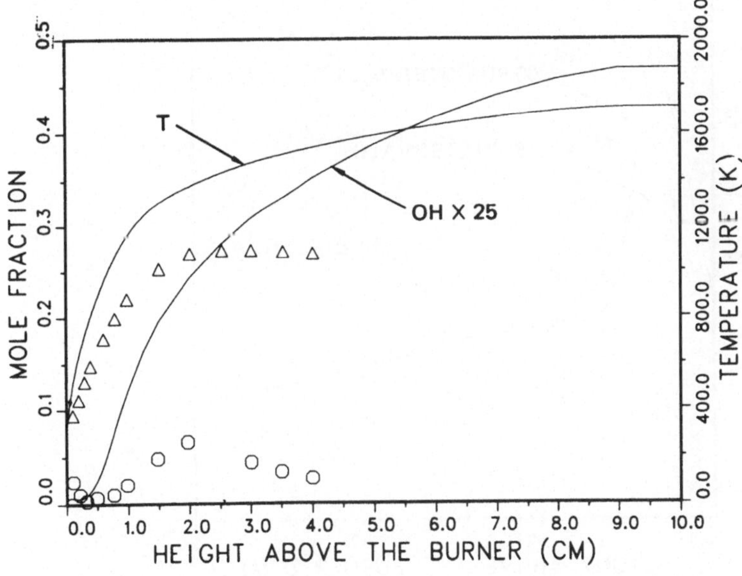

Figure 3. Experimental and calculated OH and temperature profiles (solid lines) for an H_2/O_2 flame. The calculation was performed by solving the energy and species equations together.

Methane - Oxygen Flame (CH_4/O_2)

Peters and Mahnen [25] have experimentally studied a lean, low pressure methane-oxygen flame. The experiment was performed at 40 torr and the incoming flow velocity was 67 cm/sec. The unburned mixture contained 9.5 mole percent CH_4 and 90.5 mole percent O_2. The reaction mechanism was identical to the one used by Tsatsaronis [26] in his calculations of methane-oxygen flames.

Figure 4 illustrates a family of curves representing the H_2 solution profile for the methane-oxygen system on successively finer grids. The solution on the coarsest grid (five subintervals) does not resolve the height of the peak and the profile is broadened due to numerical diffusion (compare the location of the peaks on the coarse and the fine grids). As the grid is refined, however, we secure better and better resolution of the H_2 peak until a highly resolved solution is obtained.

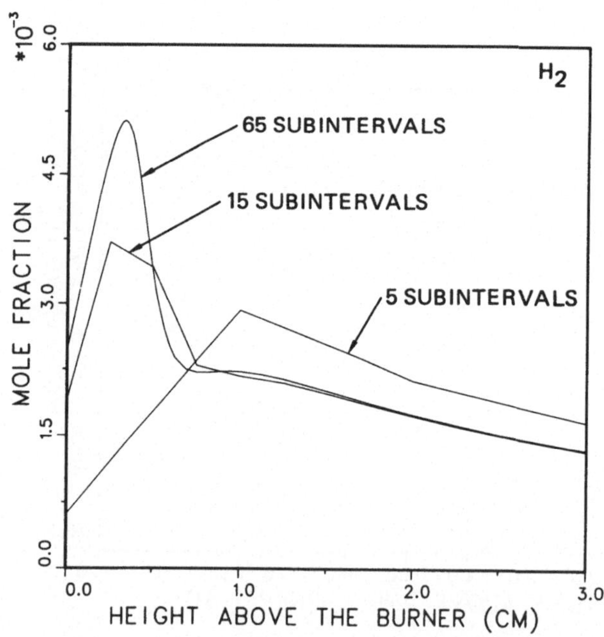

Figure 4. Calculated H_2 profiles for a methane-oxygen flame studied experimentally by Peters and Mahnen [25]. The curves represent the numerical solution on successively finer grids.

The calculation was performed by replacing the energy equation with the experimental temperature profile extended out to 5-cm, much as was done in the hydrogen-oxygen system. A solution was sought on an interval of 5-cm with an initial grid consisting of five equi-spaced subintervals. A total of 45 adaptively placed grid points were required to obtain three significant figures of accuracy with 24 of these points in the flame zone. The calculation took 41 seconds of CPU time on a CRAY-I; it was performed without thermal diffusion and with a constant cross sectional area.

Hydrogen-Air Flame (H_2/Air)

The final flame we want to consider is an atmospheric, stoichiometric, adiabatic hydrogen-air flame (one of the test problems considered in these proceedings). By making several modifications we can use the burner-stabilized flame model discussed above to solve adiabatic (freely-propagating) flames. We recall that in burner-stabilized problems the flow rate $\dot{M}$ is known. In adiabatic problems, however, $\dot{M}$ is an unknown quantity (an eigenvalue) which must be determined such that the governing equations and boundary conditions have a solution (trivial or not). To solve a generalized eigenvalue problem of this type, we apply the finite difference method discussed in Section 3 to Eqs. (2.1-2.8). To obtain a well posed problem we introduce an extra differential equation for the flow rate ($d\dot{M}/dx = 0$) along with an appropriate boundary condition. The choice of the boundary condition is flexible and we choose to specify the temperature at a point in the interior of the integration region. The unburned mixture contained 29.6 mole percent H_2, 14.8 mole percent O_2, and 55.6 mole percent N_2. The calculated adiabatic flame speed was 181 cm/sec.

In figures 5, 6 and 7 we plot the calculated temperature and species profiles for the hydrogen-air flame. The calculation was performed on a 1-cm region with an initial grid consisting of five equi-spaced subintervals. A total of 53

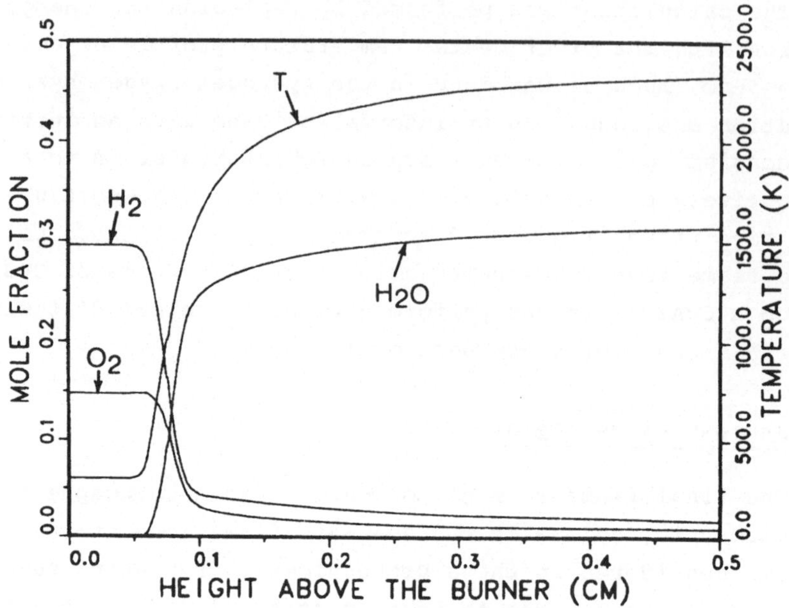

Figure 5. Calculated T, H_2, O_2 and H_2O profiles for a one atmosphere adiabatic hydrogen-air flame.

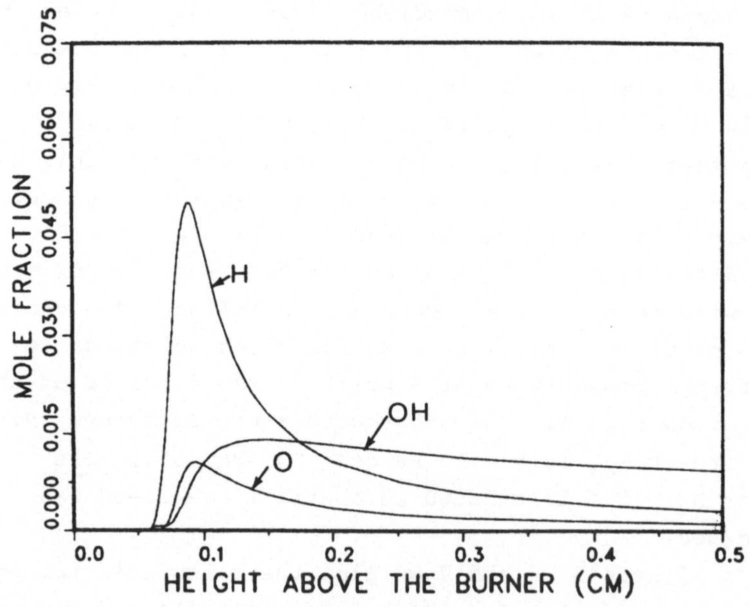

Figure 6. Calculated H, O and OH profiles for a one atmosphere adiabatic hydrogen-air flame.

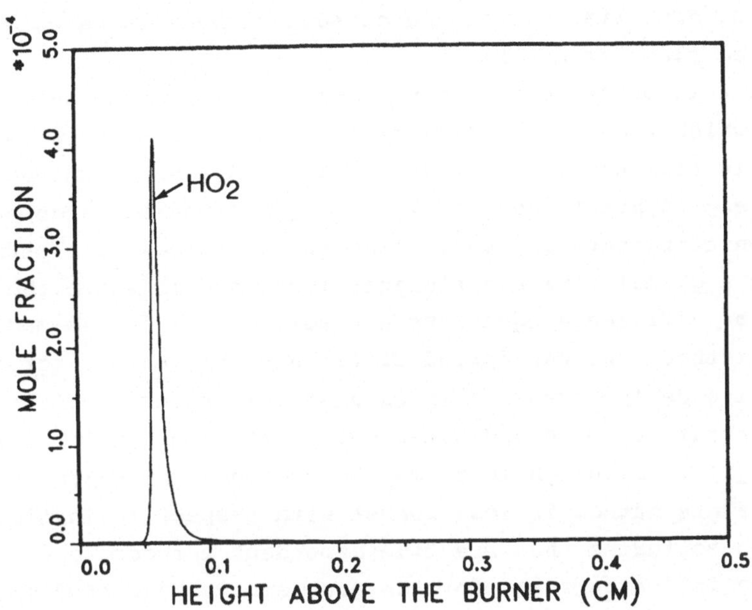

Figure 7. Calculated HO_2 profile for a one atmosphere adiabatic hydrogen-air flame.

adaptively placed grid points were used to obtain three significant figures of accuracy with 22 of these points in the flame zone. The calculation took 56 seconds of CPU time on a CRAY-I and was performed with thermal diffusion and with a constant cross sectional area.

5. Remarks

Most of the papers in the literature concerned with the numerical solution of pre-mixed flames employ some type of time dependent solution method. However, the slow relaxation of the numerical solution to steady-state can make parameter studies employing complicated transport properties extremely time consuming. Steady-state methods on the other hand, have the potential of solving the flame equations in less time than time dependent methods. Unfortunately, the often severe difficulties associated with obtaining an initial starting estimate, from which a numerical solution can be obtained, have limited the use of such methods in calculating pre-mixed flame structure.

In this paper we have developed a steady-state numerical method which reduces the sensitivity of the flame equations to the initial solution estimate. We have also shown how to place grid points adaptively in regions where the dependent solution components are undergoing rapid change. The method employs a global finite difference technique in which the nonlinear difference equations are solved by a damped-modified Newton method. We use upwind difference expressions for the convective derivatives and an assumed temperature profile to help generate a converged numerical solution on an initial coarse grid. Although there may be problems for which our steady-state method is less robust with respect to initial solution estimates than are time dependent methods, our experience indicates that, for the problems we have studied, we we have obtained reductions in CPU times by factors of 5-10 over existing time dependent solution techniques [18].

BIBLIOGRAPHY

1. D. B. Spalding, Phil. Trans. Roy. Soc. London 249A (1956).

2. G. Dixon-Lewis, Proc. Roy. Soc. London 298A (1967).

3. C. K. Westbrook and F. L. Dryer, Combust. Sci. and Tech. 20 (1979).

4. D. B. Spalding and D. L. Stephenson, Proc. Roy. Soc. London 324A (1971).

5. G. Dixon-Lewis, Proc. Roy. Soc. London 317A (1970).

6. K. A. Wilde, Combust. Flame 18 (1972).

7. R. M. Kendall and J. T. Kelly, Aerotherm TR-75-158 (1975).

8. J. O. Hirschfelder and C. F. Curtiss, J. Chem. Phys. 17 (1949).

9. C. F. Curtiss and J. O. Hirschfelder, J. Chem. Phys. 17 (1949).

10. T. P. Coffee and J. M. Heimerl, "A comparison of Transport Algorithms for Pre-Mixed, Laminar Steady-State Flames," Presented at the Fall Meeting of the Western States Section of the combustion Institute, October (1980).

11. C. R. Wilke, J. Chem. Phys. 18 (1950).

12. R. A. Svehla, NASA Technical Report R-132 (1962).

13. S. Chapman and T. G. Cowling, "The Mathematical Theory of Non-Uniform Gases," Third Edition, Cambridge University Press, Cambridge (1970).

14. P. Deuflhard, Numer. Math. 22 (1974).

15. A. B. White, Jr., SIAM J. Numer. Anal. 16 (1979).

16. V. Pereyra and E. G. Sewell, Num. Math. 23 (1975).

17. C. E. Pearson, J. Math. Phys. 47 (1968).

18. M. D. Smooke, Sandia National Laboratories Report SAND81-8040 (1982).

19. L. Monchick and E. A. Mason, J. Chem. Phys. 35 (1961).

20. JANNAF Thermochemical Tables, Dow Chemical Company, PB168370 (1965) and subsequent addenda.

21. S. Gordon and B. J. McBride, NASA SP-273 (1971).

22. R. J. Kee, J. A. Miller, and T. H. Jefferson, Sandia National Laboratories Report SAND80-8003 (1980).

23. A. C. Hindmarsh, Lawrence Livermore Laboratory Report, UCID-30150, (1977).

24. K. H. Eberius, K. Hoyermann and H. Gg. Wagner, Thirteenth Symposium (International) on Combustion, The Combustion Institute (1971).

25. J. Peters and G. Mahnen, Fourteenth Symposium (International) on Combustion, The Combustion Institute (1973).

26. G. Tsatsaronis, Combust. Flame 33 (1978).

TOWARD THE FORMULATION OF A GLOBAL LOCAL EQUILIBRIUM KINETICS MODEL FOR LAMINAR HYDROCARBON FLAMES

by

R.D. Reitz and F.V. Bracco

Princeton University

Princeton, NJ, USA

ABSTRACT

A global chemistry model for hydrocarbon combustion is formulated and applied to the computation of methane and ethane-air laminar flames. The formulation of the model is based on a first order expansion of the reaction rates about local equilibrium. The motivation for this local equilibrium model is discussed by first analysing the performance of standard global models. It is concluded that H_2 and CO need to be included in these models if the energy release is to be predicted accurately over the range of equivalence ratios of interest ($0.5 \leq \phi \leq 3$). However, the inclusion of H_2 and CO in standard irreversible reaction mechanisms is difficult. The reason is that the stoichiometric coefficients of a one-step mechanism (or the rates of the individual reactions in the equivalent multistep irreversible mechanism) have to be selected empirically or computed from the complete thermodynamic equilibrium solution, and this latter approach is computationally inefficient. A multi-step reversible reaction scheme could also be used, but for the reactions addressed in this study this requires that three Arrhenius-type forward rate expressions be specified. The proposed local equilibrium model does allow H_2 and CO and their equilibria to be included and the method is computationally efficient. The method can be applied using only one Arrhenius term and it can also be extended to include multiple time scales for the modeling of multi-step reactions. The results indicate that laminar flame speed trends such as the dependence of flame speed on equivalence ratio can be modeled satisfactorily.

INTRODUCTION

Detailed mathematical models are now being used for the combustion events in a variety of practical combustors. In these applications the modeler is often faced with the task of describing and computing the coupled processes of fluid mechanics, turbulence, heat transfer and chemistry for unsteady multidimensional, multiphase flows. Due to computer time constraints, the modeler is usually forced to sacrifice detail and to invoke simplifying assumptions which still allow the important physical processes to be described, albeit in an approximate manner.

Detailed models for the chemistry of combustion are available which describe elementary reaction mechanisms. These elementary reaction steps identify the kinetic routes by which fuel is converted into products by molecular collisions. Unfortunately, hydrocarbon combustion involves a large number of chemical species and thus it is required that a large number of partial differential equations (one for each species) be solved. In addition, although progress has been made for lower order hydrocarbons (e.g. Westbrook et al. /1/), the rate constants for the elementary reactions and the elementary reactions themselves are often unknown for general hydrocarbon fuels. These difficulties have limited the use of detailed models to fundamental studies of chemical kinetic mechanisms and flame structure.

An alternative practical approach is to describe the combustion process using overall or global mechanisms. It is assumed that the elementary reaction steps can be represented by a smaller number of overall steps which have the same form as an elementary step. The models are not intended to represent precise chemical kinetic routes, but rather the goal is to predict the important physical quantities such as the rate of heat release and the concentration of the principal energy releasing species.

In this work we present such a global model which is based on an expansion of the reaction rates about local equilibrium. The model is being explored for possible use in multidimensional engine combustion computer codes. The approach adopted was to begin by attempting to predict well the one-dimensional laminar flame speed dependence on equivalence ratio (ϕ), and also the dependence on ϕ of the concentrations of the main energy releasing species and the temperature and energy in the burned gas. The constraints of simplicity (possibly one overall chemical reaction rate) and that of sufficient generality to allow the inclusion of multistep kinetics in the model in the future were considered. The progress towards these goals is discussed here.

GOVERNING EQUATIONS AND METHOD OF SOLUTION

For the purpose of developing simplified kinetics models for laminar flame computations we consider one-dimensional unconfined flame propagation. The governing equations are /2/

$$\frac{\partial}{\partial \bar{t}} \left(\sum_{k=1}^{N} Y^{(k)} \bar{h}^{(k)} \right) + \frac{\partial}{\partial \bar{x}} \left(\frac{\rho}{\rho_o} \sum_{k=1}^{N} \bar{h}^{(k)} Y^{(k)} V^{(k)} \right) = \frac{\partial}{\partial \bar{x}} \left\{ \left(\frac{\rho}{\rho_o} \right)^2 \frac{\lambda}{\rho} \frac{\partial T}{\partial \bar{x}} \right\} - \sum_{k=1}^{N} \frac{h_o^{(k)} \omega^{(k)}}{\rho} \tag{1}$$

$$\frac{\partial}{\partial \overline{t}} Y^{(k)} + \frac{\partial}{\partial \overline{x}} \left(\frac{\rho}{\rho_o} Y^{(k)} V^{(k)}\right) = \frac{\omega^{(k)}}{\rho} \quad ; \quad k = 1,\ldots,N \qquad (2)$$

Here it has been assumed that the flame Mach number is small and hence the kinetic energy terms in the energy equation (1), and the momentum equation itself, can be dropped. A Lagrangian coordinate transformation from the physical plane (x, t) to $(\overline{x}, \overline{t})$ has been used with $\overline{t} = t$ and

$$\frac{\partial \overline{x}}{\partial x} = \frac{\rho}{\rho_o} \quad ; \quad \frac{\partial \overline{x}}{\partial t} = -\frac{\rho u}{\rho_o} \qquad (3)$$

where u is the gas velocity and ρ_o is the (constant) unburned gas density. The transformation eliminates the convective terms in the untransformed equations and the steady state flame speed is invariant under the transformation. Further details of the transformation are given by Reitz /3/.

In Eq. (1) the enthalpy $h^{(k)} = \overline{h}^{(k)} + h_o^{(k)}$, where $h_o^{(k)}$ is the enthalpy of formation of species (k), T is the gas temperature and ρ is the gas density. Equations (2) are the N species conservation equations for the N chemical species having mass fractions $Y^{(k)}$ and the $\omega^{(k)}$ are the (mass) reaction rate terms whose formulation will be discussed in following sections. The diffusion velocities in Eqs. (1) and (2) are determined from the N-1 linearly independent relations /4/

$$\frac{\rho}{\rho_o} \frac{\partial}{\partial \overline{x}} X^{(k)} = \sum_{j=2}^{N} X^{(k)} \frac{X^{(j)}(V^{(j)} - V^{(k)})}{D_{kj}} \qquad (3a)$$

supplemented by the constraint condition

$$\sum_{k=1}^{N} V^{(k)} Y^{(k)} = 0 \qquad (3b)$$

where $X^{(k)}$ are the mole fractions.

The transport coefficients in Eqs. (1) and (3) are given using the Chapman-Enskog formulae with

$$D_{ij} = 1.88 \times 10^3 \sqrt{T^3} \; \mu_{ij}/P \, \sigma_{ij}^{\,2} \, \Omega_{ij}$$

where $\mu_{ij} = 1/W^{(i)} + 1/W^{(i)}$ ($W^{(i)}$ are the molecular weights), P is the pressure given by the ideal gas law and the Lennard-Jones parameters σ_{ij} and Ω_{ij} are as tabulated in Bird et al. /5/. The mixture thermal conductivity is

$$\lambda = \sum_{i=1}^{N} \left(\frac{X^{(i)} \lambda^{(i)}}{\sum_{j=1}^{N} X^{(j)} \phi_{ij}} \right) ,$$

$$\lambda^{(i)} = \mu^{(i)} (c_p^{(i)} + 5R/4W^{(i)}) ,$$

$$\mu^{(i)} = 2.67 \times 10^{-5} \sqrt{TW^{(i)}} \, / \sigma_{ii}^{\,2} \, \Omega_{ii} ,$$

$$\phi_{ij} = \frac{1}{\sqrt{8}} \left(1 + \frac{W^{(i)}}{W^{(j)}}\right)^{1/2} \left\{1 + \left(\frac{\mu^{(i)}}{\mu^{(j)}}\right)^{1/2} \left(\frac{W^{(j)}}{W^{(i)}}\right)^{1/4} \right\}^2 .$$

Finally, the specific heats $c_p^{(i)}$ and the enthalpies are given using the polynomial fits in powers of the temperature of the Gordon McBride NASA equilibrium code data base /6/.

The initial and boundary conditions were specified as follows. Ignition was achieved by setting the initial value of the solution variables equal to their thermodynamic equilibrium values at one or more points at one end of the mesh with a step transition to the specified initial unburned gas conditions elsewhere in the computational domain. All gradients and the diffusion velocities were set equal to zero at the boundaries of the domain.

A standard explicit first order accurate numerical method was used with an equally spaced mesh for simplicity. The discretization used the approximations

$$\frac{\partial u}{\partial \bar{x}} = \frac{u_i^n - u_{i-1}^n}{\Delta \bar{x}} \; ; \qquad \frac{\partial u}{\partial \bar{t}} = \frac{u_i^{n+1} - u_i^n}{\Delta \bar{t}}$$

where Δx and Δt are the mesh spacing and timestep and the sub- and superscripts refer to the discretization levels in space and time, respectively. The timestep was adjusted during the computations by monitoring the species mass fractions at each point in the domain. In general, the timestep was increased by 1% each cycle. However, if at any time level negative mass fractions were found, the timestep was reduced. Finally, the diffusion velocities in Eq. (3) were obtained using Gauss matrix inversion.

RESULTS AND DISCUSSION

In this section we formulate and apply our local equilibrium kinetics model to hydrocarbon combustion computations. However, we first discuss the motivation for the model by analysing the performance of several standard global kinetics models. For this purpose, the spatial derivatives in Eqs. (1) and (2) were dropped and the resulting O.D.E.'s were integrated numerically with various reaction rate expressions for $\omega^{(k)}$ in Eqs. (1) and (2). The steady state solution in each case is compared with the predictions from the NASA thermodynamic equilibrium code /8/. In the second part of the study the P.D.E.'s Eqs. (1) and (2) were solved numerically to compute laminar flame speeds using the local equilibrium global kinetics model.

The simplest global chemistry model is the one-step irreversible mechanism (written for straight chain hydrocarbons) which assumes complete combustion

$$C_N H_{2N+2} + \left(\frac{3N+1}{2}\right)O_2 \xrightarrow{k} NCO_2 + (N+1)H_2O \tag{4}$$

with

$$\omega_F = -W_F k \;; \quad \omega_{O_2} = -\left(\frac{3N+1}{2}\right)W_{O_2} k \;;$$

$$\omega_{CO_2} = W_{CO_2} Nk \;; \quad \omega_{H_2O} = (N+1)W_{H_2O} k \;; \quad \omega_{N_2} = 0$$

and the subscript F refers to the fuel. For example, the reaction rate used by Gupta et al. /7/ is of Arrhenius-type with

$$k = 1.5 \times 10^{8} e^{-26020/RT} [F]^{0.5} [O_2]^{0.5} . \qquad (5)$$

This and other similar rate expressions were determined by comparison with experiment for turbulent flames, over a range of operating conditions (which include the range $0.6 \leq \phi \leq 1.5$) in an engine combustion study. The rate expression proposed by Westbrook and Dryer /8/ for laminar ethane flames is

$$k = 1.1 \times 10^{12} e^{-30000/RT} [C_2H_6]^{0.1} [O_2]^{1.65} . \qquad (6)$$

The constants in Eq. (6) (which vary depending on the fuel used) were determined using the experimental stoichiometric laminar flame speed and then by requiring that the flame speed dependence on equivalence ratio and, in particular, the flammability limits, be satisfactorily modeled. They observed that the choice of exponent on the fuel concentration in Eq. (6) plays an important role in determining the rich flammability limit and that the lean limit was rather insensitive to the choice of exponents. Thus, they chose the oxygen concentration exponent such that the computed laminar flame speed dependence on pressure would match that obtained from experiment.

The dashed curves in Figs. 1 and 2 show the computed steady state temperature and the species mass fractions, respectively, for ethane-air mixtures $(0.5 \leq \phi \leq 3)$ with $P = 1$ atm. and initial temperature $T_o = 300$ K. Here, the O.D.E.'s from Eqs. (1) and (2) were integrated using the two rates of Eqs. (5) and (6). The results show that the steady state solution is independent of the reaction rate used. Also shown in the figures are the results of the NASA equilibrium code (solid curves). The NASA program solves a system of non-linear algebraic equations for the thermodynamic equilibrium solution, and over 100 species are included in the computation.

A comparison of the results of the one-step mechanism and the

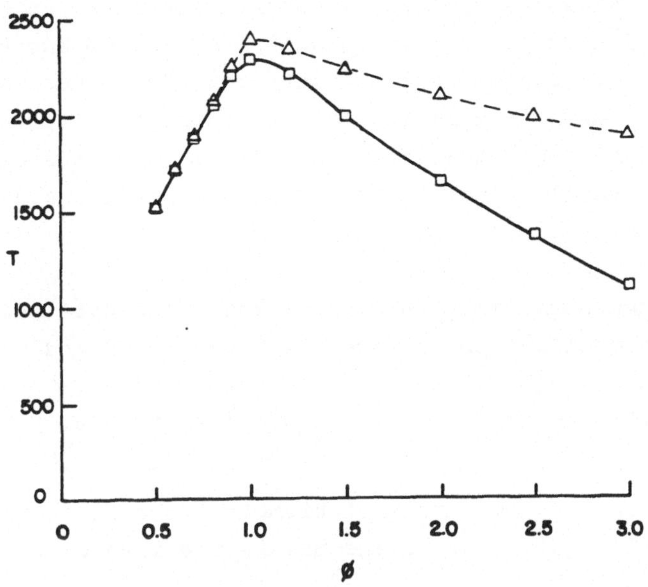

Figure 1: Adiabatic flame temperature versus equivalence ratio for ethane-air, P = 1 atm, T_o = 300 K. Dashed curve: one-step five species irreversible reaction Eq. (5) and Eq. (6). Solid curve: NASA equilibrium code. □: i) NASA code seven species, ii) one-step irreversible reaction seven species Eq. (7) and iii) local equilibrium method Eqs. (16) and (17) with rates Eq. (5) and Eq. (6).

NASA solution shows good agreement for lean mixtures, but beyond $\phi \geq 0.8$ the flame temperature and the CO_2, H_2O and fuel mass fractions are substantially overestimated with the one-step model of Eq. (4).

Also included in Figs. 1 and 2 are the results obtained using a modified version of the NASA program in which only the seven species C_2H_6, O_2, CO_2, H_2O, N_2 and CO and H_2 were included. The close agreement between the results of the complete and the modified NASA code solutions indicates that from the point of view of equilibrium energy release and concentrations of the major energy releasing species, accurate solutions can be obtained by including the two additional species, CO and H_2, over the range of equivalence ratio of interest. Parenthetically, using the full NASA program the only detectable differences were in the OH and NO concentrations near $\phi = 1$, but the concentrations were so small as not to influence the results significantly.

The findings summarized in Figs. 1 and 2 indicate that Eq. (4) should be modified to include the products CO and H_2 i.e.

$$C_NH_{2N+2} + \nu_{O_2}O_2 \xrightarrow{k} \nu_{CO_2}CO_2 + \nu_{H_2O}H_2O + \nu_{CO}CO + \nu_{H_2}H_2 \qquad (7)$$

where the ν_i are appropriate stoichiometric coefficients. Eq. (7) is formally equivalent to the multi-step irreversible mechanism

$$C_NH_{2N+2} + N/2\ O_2 \xrightarrow{k_1} NCO + (N+1)\ H_2 \qquad (8)$$

$$CO + 1/2\ O_2 \xrightarrow{k_2} CO_2 \qquad (9)$$

$$H_2 + 1/2\ O_2 \xrightarrow{k_3} H_2O \qquad (10)$$

with

$$k = k_1 \ ; \quad \nu_{CO_2} = k_2/k_1 \ ; \quad \nu_{H_2O} = k_3/k_1 \ ;$$

$$\nu_{H_2} = N+1 - \nu_{H_2O} \ ; \quad \nu_{CO} = N - \nu_{CO_2} \ ; \qquad (11)$$

$$\nu_{O_2} = (N + \nu_{CO_2} + \nu_{H_2O})/2$$

where the C-H-O atom conservation has been used in deriving Eqs. (11). Notice that in the case of the one-step mechanism Eq. (4) the <u>four</u> C-H-O-N atom balance equations specify the stoichiometric coefficient for the <u>five</u> participating species, fuel, O_2, CO_2, H_2O and inert N_2, provided that the reaction rate tends to zero as the fuel or oxygen concentrations tend to zero for $\phi > 1$ or $\phi < 1$, respectively. This requirement is satisfied by the rates in Eqs. (5) and (6). In contrast, Eq. (11) now contains two undetermined stoichiometric coefficients (νCO_2 and νH_2O) corresponding to the introduction of two additional species. In this case, the steady state solution depends on the choice of these stoichiometric coefficients or, equivalently, from Eq. (11), on the rates of the individual reactions Eqs. (8), (9) and (10).

Indeed, the dependence of the steady state products on the individual rates of a multi-step irreversible mechanism is illustrated by examining the rate proposed for reaction (9) in the four step mechanism of Hautman et al. /9/ viz.

$$k_2 = 10^{14.6} e^{-40000/RT} [CO] [O_2]^{0.25} [H_2O]^{0.5} S \qquad (12)$$

where

$$S = \min(1, 7.93 e^{-2.48\phi}) .$$

This four step mechanism is similar to that of Eqs. (8), (9) and (10) except that Eq. (8) is replaced by two steps, with the fuel being pyrolysed to an intermediate hydrocarbon species in the first step. The rate Eq. (12) with $S = 1$ was first proposed by Dryer and Glassman /10/ for lean CO oxidation. However, Hautman et al. /9/ found that this rate led to values of the rate of CO_2 formation and, correspondingly, CO_2 concentrations which were too high for progressively richer mixtures beyond $\phi = 0.84$. Accordingly, they introduced the attenuation factor S which is a decreasing function of the initial equivalence ratio. Thus it is observed that the rates of individual reaction steps need to be selected empirically so that the steady state solution approximates the thermodynamic equilibrium solution. Unfortunately, this is difficult to achieve in applications where the charge is not homogeneously distributed in space and time.

An alternative approach is to compute the stoichiometric coefficients νH_2O and νCO_2 or, equivalently to compute the rate

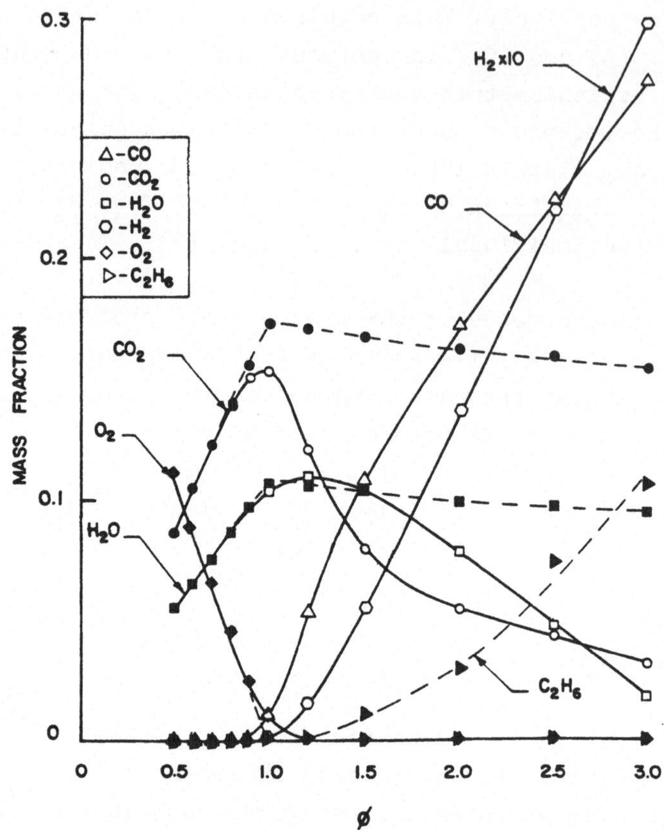

Figure 2: Burned gas mass fractions versus equivalence ratio for ethane-air, P = 1 atm, T_o = 300 K. Solid symbols: one-step five species irreversible reaction Eq. (4) with rates Eq. (5) and Eq. (6). Solid curves: NASA equilibrium code. Open symbols: i) NASA code seven species, ii) one-step irreversible reaction seven species Eq. (7) and iii) local equilibrium method Eqs. (16) and (17) with rates Eq. (5) and Eq. (6).

constants k_2 and k_3 from the thermodynamic equilibrium constants. In this case the required stoichiometric coefficients are

$$\nu_{CO_2} = N/(1 + x_{eq}) \ ,$$

$$\nu_{H_2O} = (1+N)/(1 + K_{H_2O}/K_{CO_2} \ x_{eq}) \qquad (13)$$

where x_{eq} is the CO-CO_2 mole ratio at thermodynamic equilibrium which can be found as is indicated in Appendix A.

The steady state results obtained using the seven species irreversible mechanism, Eq. (7) with the stoichiometric coefficients given by Eqs. (11) and (13) are also shown in Figs. 1 and 2 (one-step, irreversible, 7 species). The agreement between these results and those of the NASA program confirms that the proper equilibrium composition is obtained and that the inclusion of CO and H_2 is sufficient to model the energy release accurately. As expected the steady state solution was also found to be independent of the reaction rate k in Eq. (7).

However, the solution algorithm is computationally inefficient since the complete thermodynamic equilibrium solution is required at each time step (and at each space point for the case of P.D.E.'s, see Appendix A). Moreover, the method is not suited to applications where the thermodynamic state of the gas varies with space and time, since the reaction(s) are still irreversible. This deficiency can be avoided by including the reverse reactions for reactions (9) and (10). In this case the forward rate constants k_2^f and k_3^f must be specified independently. (But the reverse rates must be related to the forward rates through the equilibrium constants, thus assuming that the reaction rates away from equilibrium are equal to the rates at thermodynamic equilibrium.) However, the specification of Arrhenius-type forward rates involves the introduction of about five new empirical constants for each additional reaction (pre-exponential, power of temperature in the frequency factor, activation energy and the reaction orders). Therefore, it is of interest to seek alternative approaches which minimize the number of required input constants for modeling applications.

The local equilibrium approach to be discussed next does allow global kinetics modeling with only one Arrhenius term and equilibria which change with time are accounted for. The method follows from a first order expansion of the reaction rates about local equilibrium, i.e.

$$\frac{d}{dt}(\rho\vec{Y}) = \vec{\omega} = \vec{\omega}^* - J^*(\vec{Y} - \vec{Y}^*) \tag{14}$$

where $\vec{Y}$ and $\vec{\omega}$ are N component vectors and $\vec{\omega}^* = 0$ at local equilibrium. The Jacobian matrix in Eq. (14) is

$$J^* = -\left.\frac{\partial\omega^{(k)}}{\partial Y^{(\ell)}}\right)^*_{\rho,T} \tag{15}$$

where ρ and T are the local (and instantaneous) gas density and temperature. Notice that this formulation of the chemical rate equations is similar to that of the vibrational rate equation which is commonly used in the study of non-equilibrium gas dynamics /11/.

In principle, the Jacobian matrix can be determined for any system of elementary reactions from the given elementary reaction rates. However, for the purpose of formulating a global model, we assume that J^* is diagonal with one eigenvalue $\lambda = 1/\tau$, where τ is the single characteristic chemical reaction time scale. In this case Eq. (14) becomes

$$\frac{d}{dt}(\rho Y^{(k)}) = -\frac{1}{\tau}(Y^{(k)} - Y^{*(k)}) \tag{16}$$

and the Y^* are as given in Eqs. A-8, where the CO-CO_2 mole ratio x is again found by solving the quartic equation A-7. Notice, however, that the coefficients a_i in Eq. A-7 are evaluated using the previous time gas temperature and density so that the energy equation A-9 is not solved simultaneously. This is important since the quartic can be solved efficiently (e.g. explicitly or with a Newton polynomial algorithm) without substantial computational time penalties.

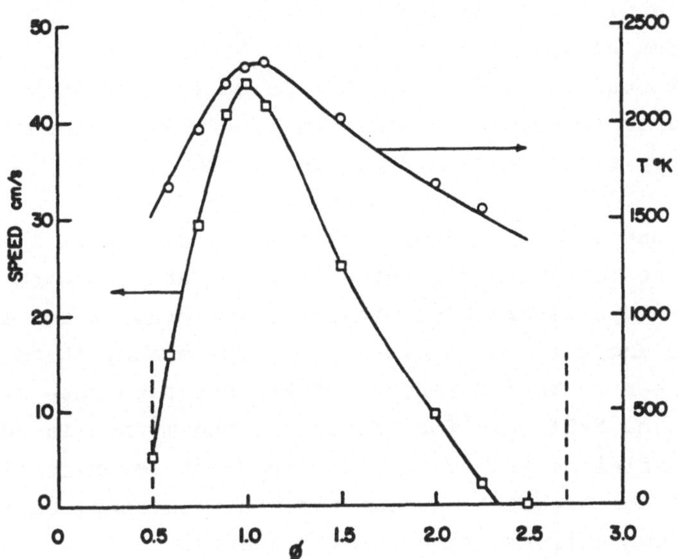

Figure 3: Laminar flame speed and burned gas temperature versus equivalence ratio using local equilibrium method, Eqs. (16) and (17) with $k = 1.5 \times 10^8 e^{-30000/RT} [F]^{0.5}[O_2]^{0.5}$. Ethane-air, P = 1 atm, T_o = 300 K. Dashed vertical lines - experimental flammability limits.

When only the five species F, O_2, CO_2, H_2O and N_2 are considered, the terms on the right hand side of Eqs. (16) are identical to the reaction rates of the five species irreversible mechanism in Eq.(4) if the chemical time scale is chosen as

$$\tau = (k\, W_F/Y_F)^{-1} \tag{17}$$

in Eqs. (16). This may be seen by excluding CO and H_2 from Eqs. (16) and by solving for Y_F^*, $Y_{O_2}^*$, $Y_{O_2}^*$ and $Y_{H_2O}^*$ using the atom balance Eqs. A-1, 2, 3 and the constraint equation A-6, followed by substitution in Eqs. (16). The advantage of this approach is that a one-step irreversible reaction rate k, which was adequate as far as equilibrium is concerned for lean hydrocarbon combustion, can be used directly in Eq. (17). Moreover, it may be possible to use the same rate expression for computations in rich mixtures (where the assumption of complete combustion is inaccurate) by simply retaining CO and H_2 as products in Eqs. (16).

This concept of extending the one-step mechanism to rich mixtures was tested using the rates Eqs. (5) and (6) with Eqs. (16) and (17) in two separate studies, and the results are also given in Figs. 1 and 2 (Local Equilibrium). The steady state solution is again seen to be independent of the reaction rate and agrees well with the NASA solution. Moreover, the method is computationally efficient and, due to the inherent reversibility of Eqs. (16), the method is suited for practical applications in spatially non-uniform, time varying charges.

Having shown that the formulation of Eqs. (16) and (17) satisfies the equilibrium limit, we now consider if it can be applied also to reproduce laminar flame speed trends. Figure 3 shows the computed laminar flame speed and burned gas temperature versus equivalence ratio for ethane-air at $P = 1$ atm and $T_o = 300$ K. The results were obtained by solving the P.D.E.'s Eqs. (1) and (2) with the reaction terms given by Eqs. (16) and (17). The rate k was as in Eq. (5) but with the activation energy increased to 30000 cal/mole. (The effect of the increased activation energy is to reduce computed stoichiometric flame speed from about 67 cm/s to 44 cm/s which is closer to the experimental value of 40-42 cm/s for ethane-air laminar flames.) The vertical dashed lines in the figure are the experimental flammability limits given in Ref. /8/. It is seen that the lean limit is predicted well and the rich limit is predicted within 10% of the experimental value. As expected the burned gas temperature is seen to agree well with the NASA solution.

The results of Fig. 4 (solid curves) correspond to the same conditions as those in Fig. 3 except that the computations used two extreme values of k in Eq. (17), viz.

$$k = 6.0 \times 10^{12} e^{-30000/RT} [O_2]^{1.7} \qquad (18)$$

which is similar to Eq. (6), and for comparison,

$$k = 1.4 \times 10^{13} e^{-30000/RT} [C_2H_6]^{1.7} \quad . \qquad (19)$$

The pre-exponential constant and the concentration exponents were chosen such that the stoichiometric laminar flame speed and pressure dependence (for $1 < P < 25$ atm) agree with the experimental values as in Ref. /8/.

The results of Fig. 4 obtained with Eqs. (18) and (19) show that the lean limit is insensitive to the form of the rate expression, but the rich limit is influenced by the fuel exponent. This trend agrees with the one-step, five species result of Westbrook and Dryer /8/. However, the present 7 species results show that the rich limit is predicted best with a larger value of fuel exponent than that used in /8/ and the limit itself is much less sensitive to the choice of exponent than found there. This is because the flame temperature plays an important role in determining the flammability limits too.

This conclusion is supported by the flame speed results obtained for methane (dashed curve in Fig. 4). Both the rich and the lean limits are seen to be modeled satisfactorily. Here the computations used the rate

$$k = 1.8 \times 10^{8} e^{-30000/RT} [CH_4]^{1.0} \qquad (20)$$

which was chosen to give the measured flame speed ($\sim$ 37 cm/s) at $\phi = 1$ and a $P^{-1/2}$ dependence of flame speed on pressure. Eq. (20) should be compared with the 5 species rate proposed by Westbrook and Dryer /8/ for methane, viz.

$$k = 8.3 \times 10^{5} e^{-30000/RT} [CH_4]^{-0.3} [O_2]^{1.3} \qquad (21)$$

They found that a negative fuel exponent was necessary to pre-

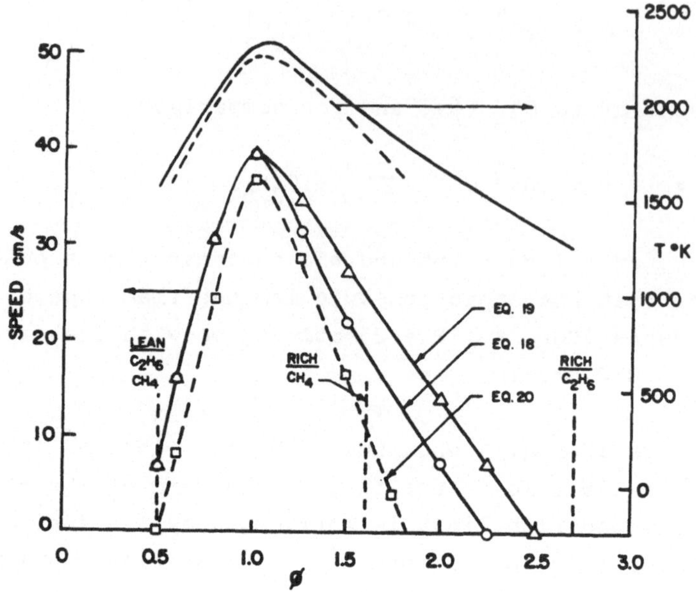

Figure 4: Laminar flame speed and burned gas temperature dependence on equivalence ratio for ethane-air and methane-air, P = 1 atm, T_o = 300 K, using local equilibrium method, Eqs. (16) and (17) with rates Eqs. (18), (19) and (20). Solid curves - ethane. Dashed curves - methane.

dict the rich flammability limit. In the present model the narrower flammability limits computed for methane compared with those for ethane are associated with the lower adiabatic flame temperature of methane (see Fig. 4). The conclusions of this study apply to the particular choice of the activation energy in Eqs. (18), (19) and (20). For example, other parametric studies have indicated that the sensitivity of the rich limit to the fuel exponent increases with a reduction in the activation energy.

Moreover, the experimental flammability limits are known to broaden with increasing pressure /12/. Therefore, the use of the stoichiometric flame speed pressure dependence to determine the exponents on the concentrations is questionable. It appears that the exponent itself should be a function of equivalence ratio (with the sum of the exponents > 2 for rich mixtures) in order to model the broadening of the flammability limits with increasing pressure.

The local equilibrium method, Eqs. (16), is currently being applied to the modeling of laminar flame propagation with higher order hydrocarbons. Results obtained with propane indicate that the influence of the choice of constants in the rate expression are similar to that of ethane. The method is also being tested using an extension of the multistep scheme Eqs. (8), (9) and (10). In this case three characteristic chemical times are included in the formulation. Finally, the influence of the particular formulation of the present local equilibrium model is being examined by comparison with results from another local equilibrium model in which the species concentrations themselves are given by their local equilibrium values. In this case the reaction proceeds by adding fuel or oxidizer at a specified rate.

CONCLUSIONS

We have shown that the further addition of only H_2 and CO in the formulation of conventional global chemical kinetics models is sufficient to allow the energy release to be predicted accurately over the range of equivalence ratios of practical interest in hydrocarbon flame propagation. This study also points out that the inclusion of these two additional species is difficult within the framework of the usual one-step irreversible reaction mechanism. This difficulty is associated with the fact that the stoichiometric coefficients, or the individual rates in the equivalent multi-step irreversible mechanism, have to be modeled empirically in order that the results approximate the thermodynamic equilibrium solution. In principle, a multi-step reversible reaction scheme would resolve these difficulties. However, for the reactions addressed here this would require the specification of three Arrhenius-type forward rate expressions (up to 15 empirical input constants) which, considering the limitations already inherent in global chemistry models, may be difficult to justify.

The local equilibrium approach presented here does allow H_2 and CO and their equilibria to be included and thus the energy release is predicted accurately. The method can be applied using only one Arrhenius term and it is computationally efficient. In addition, results for laminar flame propagation indicate that the dependence of the flame speed on equivalence ratio can be modeled satisfactorily. Finally, the method can be extended to model multi-step or elementary reaction mechanisms.

ACKNOWLEDGEMENTS

The authors wish to acknowledge the financial support of the Deutsche Forschungsgemeinschaft during the GAMM workshop. The work and travel expenses was supported by DOE, NSF, Volkswagen, FIAT and General Motors.

APPENDIX A

The algebraic equations for the thermodynamic equilibrium concentrations of the 6 reactive species consist of the 3 atom balance equations (C-H-O) and 3 non-linear equilibrium relations. For simplicity the equilibrium fuel concentration is assumed to be zero. This assumption is motivated by the NASA code prediction of negligible unburned fuel for $\phi < 3$ which includes the range of interest in this work. (For $\phi > 3$ unburned hydrocarbons such as CH_4 do appear among the equilibrium products.) For straight-chain hydrocarbons $F = C_N H_{2N+2}$ the equations are

$$Nn^*_F + n^*_{CO_2} + n^*_{CO} = Nn_F + n_{CO_2} + n_{CO} = \alpha_C \quad \text{(A-1)}$$

$$(2N+2)n^*_F + 2n^*_{H_2} + 2n^*_{H_2O} = (2N+2)n_F + 2n_{H_2} + n_{H_2O} = \alpha_H \quad \text{(A-2)}$$

$$2n^*_{O_2} + n^*_{CO} + 2n^*_{CO_2} + n_{H_2O} = 2n_{O_2} + n_{CO} + 2n_{CO_2} + n_{H_2O} = \alpha_O \quad \text{(A-3)}$$

$$K_{CO_2}(T) = n^*_{CO_2}/n^*_{CO_2}(n^*_{O_2}\rho)^{1/2} \quad \text{(A-4)}$$

$$K_{H_2O}(T) = n^*_{H_2O}/n^*_{H_2}(n^*_{O_2}\rho)^{1/2} \quad \text{(A-5)}$$

$$n^*_F = 0 \quad \text{(A-6)}$$

where $n_i = Y^{(i)}/W^{(i)}$ is the molar concentration and * indicates local or thermodynamic equilibrium values depending on whether the temperature and density in Eqs. A-5, 6 corresponds to their local or their thermodynamic equilibrium values. The equilibrium constants were found from least squarefits to the JANAF Table thermochemical data as

$$K_{CO_2} = 5.86 \times 10^{-16}\, e^{68354/RT}\, T^{0.99}$$

$$K_{H_2O} = 0.68\, e^{57938/RT} \quad .$$

Defining $x = n^*_{CO}/n^*_{CO_2}$ and rearranging Eqs. A-1 - A-6 gives

$$a_4x^4 + a_3x^3 + a_2x^2 + a_1x + a_o = 0 \quad \text{(A-7)}$$

where

$$a_4 = K(\alpha_c - \alpha_o) \ ; \quad a_3 = (1+2K)\alpha_c - (1+K)\alpha_o + \alpha_H/2 \ ;$$

$$a_2 = 2\alpha_c - \alpha_o + \alpha_H/2 + \alpha_o K \quad ; \quad a_1 = \alpha_o(1+K) \; ;$$

$$a_o = 2/\rho K^2_{CO_2} \quad \text{and} \quad K = K_{CO_2}/K_{H_2O} \; .$$

The equilibrium mass fractions are given by

$$Y^*_{CO_2} = W_{CO_2}\alpha_c/(1+x) \; ; \qquad Y^*_{H_2O} = W_{H_2O}\alpha_H/2(1+kx) \; ;$$

$$Y^*_{CO} = W_{CO}x\alpha_c/(1+x) \; ; \qquad Y^*_{H_2} = W_{H_2}\alpha_H Kx/2(1+kx) \; ;$$

$$Y^*_{O_2} = W_{O_2}/\rho x^2 K^2_{CO_2} \; ; \qquad Y^*_F = 0 \tag{A-8}$$

In order to find the thermodynamic equilibrium solution, Eq. (A-7) is coupled with the energy equation

$$\sum_{reac.} Y^{(k)} \left(h_o^{(k)} + \int_{T_o}^{T_{reac}} cp^{(k)} dT'\right) = \sum_{prod.} Y^{(k)}_{eq} \left(h_o^{(k)} + \int_{T_o}^{T_{eq}} cp^{(k)} dT'\right) \tag{A-9}$$

through the temperature dependence of the coefficients a_i. At thermodynamic equilibrium the CO-CO_2 mole ratio $x = x_{eq}$, the temperature is the adiabatic flame temperature T_{eq} and the *'s in Eq. (A-8) are replaced by the subscript eq.

REFERENCES

/1/ C.K. Westbrook, J. Creighton, C. Lund and F.L. Dryer, "A Numerical Model of Chemical Kinetics of Combustion in a Turbulent Flow Reactor", J. Phys. Chem. 81 (1977), pp. 2542-2554.

/2/ R.D. Reitz, "Computations of Laminar Flame Propagation Using an Explicit Numerical Method", 18th Symposium (Int.) on Combustion, The Combustion Institute (1980), 433-442.

/3/ R.D. Reitz, "The Application of an Explicit Numerical Method to a Reaction-Diffusion System in Combustion", Report COO-3077-162, Courant Mathematics and Computing Laboratory, New York University, NY, 1979.

/4/ F.A. Williams, "Combustion Theory", Addison-Wesley, 1965.

/5/ R.B. Bird, W.E. Stewart and E.N. Lightfoot, "Transport Phenomena", Wiley, New York, 1960.

/6/ S. Gordon and B.J. McBride, "Computer Program for Calculation of Complex Chemical Equilibrium Compositions Rocket Performance, Incident and Reflected Shocks and Chapman-Jouget Detonations", NASA SP-273 (1971).

/7/ H.C. Gupta, R.L. Steinberger and F.V. Bracco, "Combustion in a Divided Chamber, Stratified Charge, Reciprocating Engine: Initial Comparisons of Calculated and Measured Flame Propagation", Comb. Sci. and Techn. 22, (1980), pp. 27-61.

/8/ C.K. Westbrook and F.L. Dryer, "Simplified Reaction Mechanisms for the Oxidation of Hydrocarbon Fuels in Flames", Lawrence Livermore Laboratory Report UCRL-84943, 1980.

/9/ D.J. Hautman, F.L. Dryer, K.P. Schug and I. Glassman, "A Multiple-Step Overall Kinetic Mechanism for the Oxidation of Hydrocarbons", Comb. Sci. and Techn. 25, (1981), pp. 219-235.

/10/ F.L. Dryer and I. Glassman, "High-Temperature Oxidation of CO and CH_4", 14th Symposium (International) on Combustion, The Combustion Institute, (1973), p. 987.

/11/ W.G. Vincenti and C.H. Kruger, "Introduction to Physical Gas Dynamics", Wiley, New York, 1967.

/12/ B. Lewis and G. von Elbe, "Combustion, Flames and Explosions of Gases", Academic Press, New York, 1961.

Time-Dependent Simulation of Flames in Hydrogen-Oxygen-Nitrogen Mixtures

K. Kailasanath*, E.S. Oran, J.P. Boris, and T.R. Young
Laboratory for Computational Physics
Naval Research Laboratory
Washington, D.C., 20375

1. INTRODUCTION

This paper describes a one-dimensional, time-dependent, Lagrangian model developed to study the initiation, propagation and quenching of laminar flames. The model incorporates a number of new approaches and algorithms which have now been tested by comparisons to less complex or analytic solutions and by comparisons to experimental data. These new elements include: ADINC [1] an implicit, Lagrangian method for solving the convective parts of the conservation equations; DFLUX [2,3] a variable accuracy algorithm for determining diffusion fluxes without having to invert matrices; SPLIT and MERGE, routines for dividing or merging computational cells as specified by external criteria; VSAIM, a vectorized version of the ordinary differential equation solver, CHEMEQ [4,5]; and a new method for treating an open boundary in an implicit, Lagrangian calculation. An asymptotic coupling method, used in conjunction with timestep splitting to couple the various processes, allows the use of entirely different algorithms for the physical processes represented by different mathematical forms.

The model has been used for a variety of flame studies of hydrogen-oxygen-nitrogen mixtures. These include calculations of minimum ignition energies, flammability limits, quench volumes, and burning velocities. The chemical rate scheme has now been tested extensively, as have the thermal and molecular diffusion coefficients. Thus we expect the model to calculate correctly the time-dependent behavior implied by the initial and boundary conditions supplied.

*Currently with Science Application, Inc. McLean, VA

After a summary is given of the important numerical features, several calculations are presented. The first is a flame initiation and minimum ignition energy study of a mixture of $H_2:O_2:N_2/2:1:10$ initially at 298 K and 1 atm. These are inherently time-dependent problems and take advantage of this property of the model. The second problem discussed is a calculation of the burning velocity of the mixture $H_2:O_2:N_2/2:1:4$, again at 298 K and 1 atm. The burning velocity is a quantity used to describe a steady state property and it is generally calculated by steady state methods. We show below that the model described also does well on this type of calculation.

2. NUMERICAL MODEL

2.1 Basic Equations

We solve the time-dependent equations for conservation of total mass density ρ, momentum $\rho\underline{v}$, and energy E as well as the individual species number densities $\{n_j\}$. These may be written as [2,6]:

$$\frac{\partial\rho}{\partial t} = -\underline{\nabla}\cdot\rho\underline{v} \tag{1}$$

$$\frac{\partial n_j}{\partial t} = -\underline{\nabla}\cdot n_j\underline{V}_j - \underline{\nabla}\cdot n_j\underline{v} + P_j - L_j n_j \tag{2}$$

$$\frac{\partial\rho\underline{v}}{\partial t} = -\underline{\nabla}\cdot(\rho\underline{v}\underline{v}) - \underline{\nabla} P + \underline{\nabla}\cdot\eta_m[\nabla v + (\underline{\nabla v})^T] \tag{3}$$

$$\frac{\partial E}{\partial t} = -\underline{\nabla}\cdot E\underline{v} - \underline{\nabla}\cdot P\underline{v} - \underline{\nabla}\cdot\underline{Q} \tag{4}$$

where the heat flux vector, $\underline{Q}$, is defined as

$$\underline{Q} = -\lambda_m\underline{\nabla}T + \sum_j n_j h_j \underline{V}_j + k_B T \sum_{j,k}\frac{n_j D_j^T}{NM_j D_{jk}}(\underline{V}_j - \underline{V}_k) \tag{5}$$

The quantity $\underline{v}$ is the fluid velocity, the $\{\underline{V}_j\}$ are the diffusion velocities, and $\{P_j\}$ and $\{L_j\}$ refer to the chemical production and loss processes for the individual species j. The quanties η_m and λ_m are the mixture viscosities and thermal conductivities respectively. The superscript "T" in the last term of Eq. (3) indicates that the transpose is taken. The quantities $\{h_j\}$ are the temperature dependent enthalpies for each species, and the $\{D_{jk}\}$ and $\{D^T{}_j\}$ are the sets of binary and thermal diffusion coefficients, respectively.

We also assume that the mixture consists of ideal gases so that the pressure, P, may be written as:

$$P = Nk_BT \tag{6}$$

where N is the total number density, k_B is Boltzmann's constant and T is the temperature. The model presented in this report, however, is not restricted to ideal gases and in fact any equation of state may be used.

The diffusion equations may be written as

$$S_j = \sum_k \frac{n_j n_k}{N^2 D_{jk}} (\underline{V}_k - \underline{V}_j) \tag{7}$$

where the source terms S_j are defined as

$$S_j \equiv \underline{\nabla}(\frac{n_j}{N}) - (\frac{\rho_j}{\rho} - \frac{n_j}{N}) \frac{\underline{\nabla P}}{P} - \sum_k \frac{n_j n_k}{N^2 D_{jk}} (\frac{D_k^T}{\rho_k} - \frac{D_j^T}{\rho_j}) \frac{\underline{\nabla T}}{T} \tag{8}$$

The diffusion velocities are also subject to the constraint:

$$\sum_j \rho_j V_j = 0 \tag{9}$$

2.2 Convective Transport

The convective transport terms in Eqs. (1-4) are solved by the algorithm ADINC [1] . ADINC, designed for either Adiabatic or incompressible flows, is an implicit and Lagrangian algorithm. Since it communicates compression and expansion across the system implicitly, it overcomes the Courant time-step limit. Since it is Lagrangian, it can maintain steep gradients computationally for a long period of time. This is important in flame calculations where the diffusive transport of material and energy can govern the system evolution and therefore must be calculated accurately.

ADINC solves the following equations for mass and momentum transport in one dimension:

$$\frac{d\rho}{dt} = - \rho \, \underline{\nabla} \cdot \underline{v} \tag{10}$$

$$\rho \frac{d\underline{v}}{dt} = - \underline{\nabla P} \qquad (11)$$

The energy evolution equation is eliminated by using an adiabatic equation of state:

$$\rho(P,S) = \rho_c + (P/S)^{1/\gamma} \qquad (12)$$

This equation of state with ρ_c = 0 is correct for adiabatic compression and expansion of an ideal gas. The entropy, S, is assumed constant throughout the numerical integration. Non-adiabatic processes, such as external heating, thermal conduction and chemical energy release, are added to Eqs. (10-11) using the timestep splitting methods described below.

Given an aproximation to the pressure, ADINC calculates the fluid density using Eq. (12). This equation-of-state density is compared to the density derived from the fluid dynamics through Eq. (10). The difference is iterated to zero using a quadratically convergent implicit solution of Eq. (11) which then gives an improved approximation to the pressure. During this iteration that analytic derivative $d\Lambda/dP$ is used where Λ is the volume of a computational cell. Thus

$$\frac{1}{\Lambda}\frac{d\Lambda}{dP} = - \frac{1}{\gamma\rho P} (P/S)^{1/\gamma} \qquad (13)$$

for the particular equation of state (12). ADINC also assumes that pressure and density are constant within each individual finite-difference cell and that the physics is evolving slowly enough for full communication across that cell to have occurred in a timestep.

ADINC has been used extensively for solving a wide variety of problems. Some of these and a number of tests of the algorithm have been documented by Boris [1].

2.3 Chemical Kinetics Calculations

The coupled, nonlinear, ordinary differential equations which describe the chemical interactions are taken from that part of Eq. (2), which represents the production and loss of reacting species:

$$\frac{\partial n_j}{\partial t} = P_j - L_j n_j \text{ , } j = 1, \ldots M, \qquad (14)$$

where M is the total number of species present. The functional dependencies of the terms

$$P_j = P_j\{n_k(t)\}$$

$$L_j = L_j\{n_k(t)\} \quad , \; k = 1, \ldots., M \tag{15}$$

emphasize the strong coupling between the various species. The system represented by Eq. (14) may be stiff when there are large differences in the time constants associated with different chemical reations. Stiffness may occur for different species, in different locations, at different times or simultaneously throughout the course of an integration. Because of this, when there are a large number of reactions, the solution of the chemical kinetics equations is usually the most expensive part of a detailed reactive flow calculation. Furthermore, the computational cost increases with the number of species and the dimensionality of the problem. Therefore a method which is efficient, accurate, conservative, stable and which does not require storage of large quantities of data from one timestep to another is required. Such a method is VSAIM, which is a fully vectorized version of the selected asymptotic integration method employed in CHEMEQ [4,5]. In this method, the stiff equations are identified and solved using a very stable asymptotic method. The remaining equations are solved using a standard classical method.

Once the species number densities are known, the temperature can be evaluated by iterating on the equation:

$$E_T - \frac{1}{2}\rho v^2 = \sum_j h_{o_j} n_j + \sum_j h_j n_j - P \tag{16}$$

where E_T is the total energy per unit volume, h_{oj} is the set of heats of formation of the species j and h_j is the sensible enthalpy of species j. A Newton-Raphson iteration technique is used and it usually converges in two or three iterations per timestep.

2.4 Diffusive Transport

The diffusive transport processes considered in this model are molecular diffusion and thermal conduction. These are the parts of Eqs. (1-4) which are represented by the expressions:

$$\frac{\partial n_j}{\partial t} = - \underline{\nabla} . n_j \underline{V}_j \tag{17}$$

$$\frac{\partial E}{\partial t} = \underline{\nabla} . [-\lambda \underline{\nabla} T + \sum_j h_j n_j \underline{V}_j]. \tag{18}$$

The above equations are conservatively differenced and solved explicitly. The effects of viscosity and thermal diffusion have not been considered for the results discussed in this paper.

The diffusion velocities $\{V_j\}$ in Eqs. (17, 18) are determined by solving the equations [1,2]

$$S_j = \sum_{\substack{k=1 \\ k \neq j}}^{M} \frac{n_j n_k}{N^2 D_{jk}} (\underline{V}_k - \underline{V}_j) = \underline{\nabla} (n_j/N) - (\rho_j/\rho - n/N) \frac{\underline{\nabla} P}{P} \tag{19}$$

subject to the constraints

$$\sum_{j=1}^{M} S_j = 0, \text{ and } \sum_{j=1}^{M} \rho_j V_j = 0. \tag{20}$$

An iterative algorithm for solving for the diffusion velocities which avoids the cost of performing matrix inversions has been developed [2,3]. This algorithm DFLUX, is of $O(M^2)$ and is vectorized. Thus it is substantially faster than $O(M^3)$ matrix inversions when four or more species are involved.

Equations for the evaluation of diffusive transport coefficients have been discussed in detail by Picone [7] and summarized in Oran and Boris [2]. The forms given are for mixtures of neutral gases. Their derivations are based on or are advancements of the fundamental work of Chapman and Cowling [8] and Hirschfelder, Curtis and Bird [9]. Representations of the coefficients which are easily used in reactive flow models and not requiring the expensive inversion of matrices are chosen.

2.5 Timestep Splitting

In the asymptotic timestep split approach [2], the individual terms in the Eqs. (1-4) are solved independently as described above and then asymptotically coupled together. Since both chemistry and sound waves are usually stiff in deflagration problems, special care is required in coupling the chemical heat release to the fluid dynamics. In a flame, fluid dynamic expansion and diffusive transport relieve the pressure from the flame region as fast as it is generated. Thus the pressure stays effectively constant. Small pressure fluctuations, $O(\underline{v}^2/c^2)$, do exist and are just large enough to drive the flows which reapportion the energy released by chemistry or transported by diffusion. The chemistry step should be taken at constant pressure, but it may also be taken at constant volume with temperature held fixed if the profiles only change slowly per timestep. At the completion of the chemistry integration, the heat release is converted to an effective pressure change at constant volume because the cell volume has been held fixed during the chemical kinetics calculation. This pressure change is then used as an energy source term for the fluid dynamics timestep to get the correct fluid expansion. This is done by modifying the cell entropies used as the fluid dynamics module, ADINC, described earlier. At this point thermal conduction and diffusion heat fluxes also contribute changes to the entropy.

2.6 Open Boundary Condition

In the study of unconfined flames, a representation for the open-boundary is required since the size of the computational domain is limited. One approach is to allow the computational cells to increase in size as they get further from the flame. Thus the computational domain is very large and there is no corresponding increase in computer storage. However, the cell stretching should be gradual to limit inaccuracies which arise as a result of the varying cell size.

A different approach used in the study of the propagation of unconfined flames is discussed below. Here the effects of an open boundary are accounted for by allowing the pressure changes which occur in each cell during the timestep to relax adiabatically to the pressure before the time step. Due to this relaxation, the volume of the cell changes since a Lagrangian coordinate system has been used. The changes in the volume of the cells causes the location of the open boundary to change. The location of the open boundary and the fluid velocity in the last cell (which is also the velocity of the open boundary) are used as open boundary conditions. This procedure is appropriate because in unconfined flames the pressure stays effectively constant.

3. FLAMES IN HYDROGEN-OXYGEN-NITROGEN MIXTURES

Simulations of flames in hydrogen-oxygen-nitrogen mixtures have been carried out using the numerical model described above. Applying the model to a specific gas mixture requires knowledge of the chemical kinetic rate scheme and other species dependent input parameters such as thermal conducitvity and molecular diffusion coefficients.

The chemical kinetics rate scheme used consists of about fifty chemical rates relating the species H_2, O_2, H, O, OH, H_2O, HO_2 and H_2O_2. It has been extensively tested against experimental data [10, 11] and shown to give good results. Burks and Oran [10] showed that the results computed with the scheme compared very well with experimentally observed induction times, second explosion limits and the temporal behavior of reactive species. Oran et al. [11] have shown that the scheme gives good results when coupled with a fluid dynamic model in the simulation of the conditions behind a reflected shock. The reaction rate scheme has not been presented here since it is readily available [10,11,12]. Heats of formation and enthalpies have been taken from the JANAF tables [13].

The diffusion coefficients have been obtained using the data given by Marrero and Mason [14]. The Lennard-Jones (12:6) potential parameters σ and (ε/k) are those given by Svehla [15] except for σ_H and σ_{H_2O} for which the values given by Dixon-Lewis [16] have been used.

Below we first discuss some of the results obtained in a study of the ignition and quenching of a $2:1:10/H_2:O_2:N_2$

mixture. Then we present results from a simulation of a propagating flame in a H_2-air mixture. In both cases, the initial temperatures and pressures of the unburned gas were 298 K and 1 atm, respectively.

3.1 Minimum Ignition Energies and Quenching Distances

The model described above is ideally suited for studying time-dependent problems such as the ignition of gas mixtures [12]. The model was configured in spherical geometry with an open boundary at one end to simulate a very large system. Energy was deposited in the center with a radius of deposition, R_o. Results from a typical calculation are presented in Fig. 1. The figure depicts the time history of the temperature profile after 4 mJ of energy is deposited over a period of 0.1 ms. Even after the energy deposition is stopped, the central temperature continues to increase due to the heat released in chemical reactions. With time, however, the temperature near the center decreases and the temperature away from the center increases due to diffusive transport. By 4.5 ms, we see that the temperature distribution exhibits a "flame" temperature profile. If the amount of energy deposited, E_o, was reduced to 3 mJ, the temperature distribution does not develop into a flame temperature profile.

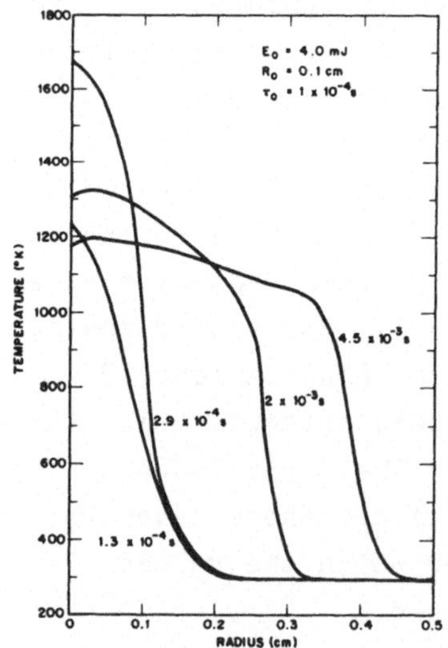

Fig.1. Time history of the temperature profile.

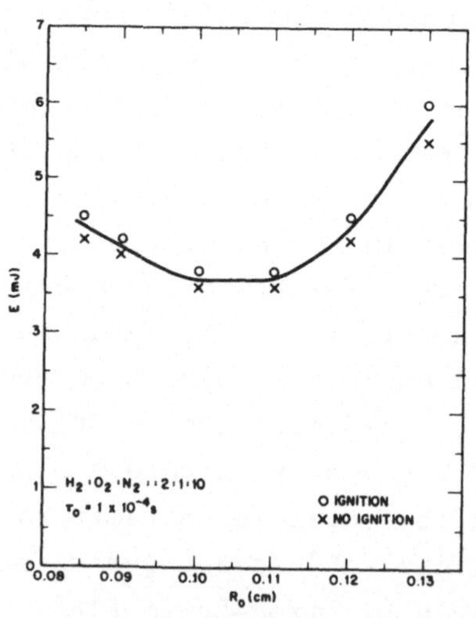

Fig.2. Minimum ignition energy as a function of radius of energy deposition.

By repeating the computations for different values of E_0, a bound for the minimum ignition energy for that particular radius was obtained. Similar calculations were performed for different values of the radius of deposition, R_0. The results of such investigations are shown in Fig 2. A propagating flame results when 3.8 mJ of energy is deposited in a sphere with a radius of 0.1 cm. However if the same amount of energy is deposited in a sphere of smaller radius, the rate of heat liberation is insufficient to compensate for the rate of heat loss and consequently there is no ignition. This radius, 0.1 cm, is the "quench-radius" for this particular mixture. For radii slightly larger than the quench-radius, the minimum ignition energy is almost constant and for larger radii (larger than 0.11 cm) the minimum ignition energy increases rapidly with increasing radii. Therefore for the system under study, the absolute minimum energy is about 3.7 mJ. These observations are in qualitative agreement with those of Lewis and von Elbe [17]. Quantitative comparisons are not possible since the composition of the mixture and the time for energy deposition are different.

3.2 Calculation of the Burning Velocity of an H_2:O_2:N_2/2:1:4 Mixture at 298K, 1 atm.

The model described above can also be used to study the propagation of flames in pre-mixed gas mixtures. As shown in Fig. 1, if sufficient energy is deposited at the center of a sphere and computations are carried out for a sufficiently long time, the temperature distribution attains a typical "flame profile". However, this method is an expensive way to generate a propagating flame since so much time is spent in establishing and overcoming the initial conditions. Another method for initializing the problem quickly is to start the computations with a good guess for the temperature and species profile in a region behind a flame front. The closer the initial profiles are to the final, steady state flame profile, the sooner the initial conditions relax to the steady propagating flame. This procedure was adopted for obtaining the values for a flame velocity.

The model was set up in cartesian geometry with one end closed (preventing any gas flow through it) and the other open to the atmosphere. At the closed end, a first approximation to the temperature and major species profile was set up as initial conditions. This soon evolves into a propagating flame as can be seen in Fig. 3, where the temperature profiles at 50 μs and 60 μs (from the start of the computations) are shown. However, in order to determine the flame speed, a criterion for the location of the flame is required. An arbitrary value of 900K is chosen to define the location of the flame. The movement of the location of this value (900K) in time and space gives the rate of propagation of the flame. The location of the flame is presented as a function of time in curve (a) in Fig. 4. The slope of this curve gives the flame velocity (for an inertial observer). From this figure we see that the flame propagates initially at a velocity of 8.7 m/s and by 50 μs (from the start of the computations) it attains a nearly constant value of 9.7 m/s.

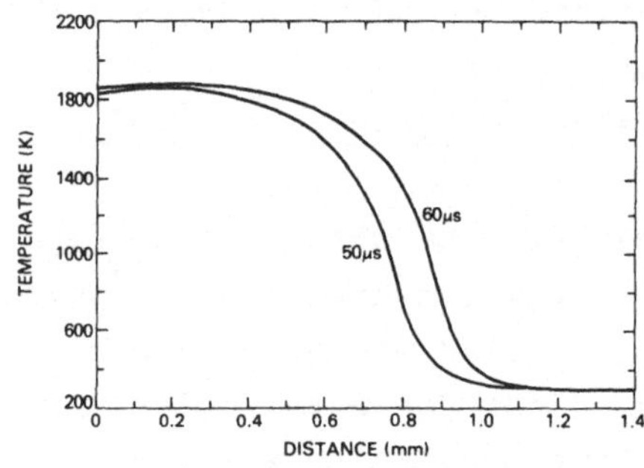

Fig.3. Temperature profiles in a propagating flame.

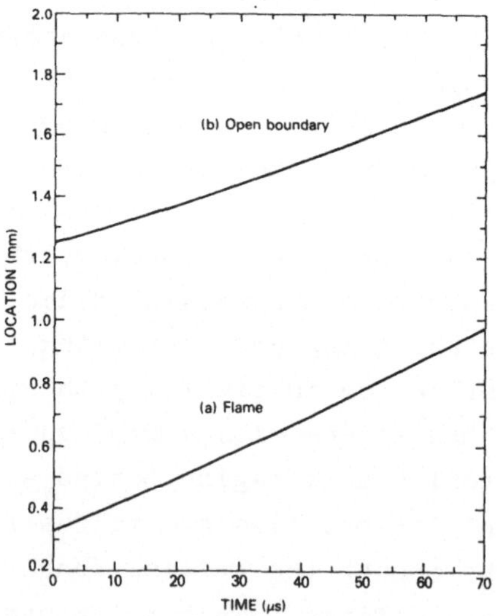

Fig.4. Time history of the location of (a) the flame and (b) the open boundary.

A quantity of considerable practical interest is the burning velocity which is defined as the rate at which the flame consumes the reactants, or equally well as the flame velocity relative to the unburned gases. In Fig. 5 the fluid velocity profile in the system is depicted at a particular time. The velocity is zero at one end since that end is a closed boundary. The flow velocity increases across the flame and is maximum at the open boundary. An estimate for the burning velocity can be obtained by subtracting the velocity of the unburned gases from the flame velocity. An upper estimate for the velocity of the unburned gases is the fluid velocity at the open boundary. Since this model uses a Lagrangian coordinate system, the fluid velocity at the open boundary is also given by the rate of movement of the open boundary. The location of the open boundary has been depicted as a function of time in Fig. 4 b . Corresponding to a flame velocity of 9.7 m/s we see that the velocity of the unburned gases (open-boundary) is about 7.6 m/s giving a burning velocity of 2.1 m/s for the hydrogen-air flame studied. This compares well with the experimental values for the burning velocity [18] which are between 2 and 2.5 m/s.

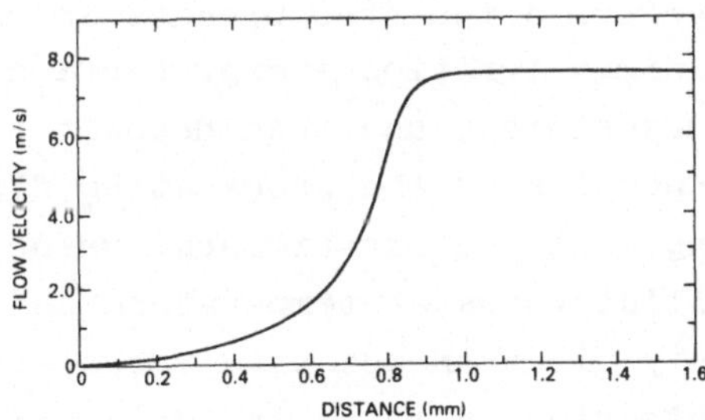

Fig.5. Flow velocity profile in a propagating flame.

IV. CONCLUSIONS

In this paper we have shown that the time-dependent model described does a good job in predicting both transient and steady state properties of laminar hydrogen-oxygen-nitrogen flames. In the calculations of the burning velocity of a mixture of $H_2:O_2:N_2/2:1:4$ at 1 atm and 298 K, the velocity predicted was at the lower end of the observed range. We note, however, that there are several factors which would tend to increase this estimated value. First, the calculated burning velocity would be larger if the velocity of the unburned gas used in this calculation were closer to the flame front instead of at the open boundary. It is not clear where it should be evaluated. Second, if thermal diffusion is included, it would increase the flame velocity by perhaps 5-10%. Finally, the extremely important hydrogen diffusion coefficients are not really known at high temperatures. If they are higher than the ones used here, the flame velocity would increase. These observations lead us to believe that the higher observed values of the burning velocity are probably more accurate.

Finally, we wish to note that the method is sensitive to initial conditions. As in the steady-state calculations, a better initial guess leads to faster convergence to a steady propagating flame. Perhaps the best way to initialize such a problem is to perform a two step process in which we first determine the minimum ignition energy for a given radius and time of energy deposition, and then deposit this energy and evaluate the properties of the propagating flamefront. However, considering that the experimental velocities vary by ~25%, a good initial guess at temperature and species concentrations certainly gives an adequate calculation. Finally, we note that the criterion on which to judge the initial guess values is to look at how flat the temperature profile is behind the flamefront. A bad guess produces some structure and oscillations which only slowly disappear.

ACKNOWLEDGEMENTS

We wish to thank Dr.J.M. Picone for many useful conversations and insights. This work was sponsored by the Office of Naval Research through the Naval Research Laboratory and by the Naval Material Command.

REFERENCES

1. Boris, J.P., ADINC: An Implicit Lagrangian Hydrodynamics Code, N.R.L. Memorandum Report 4022, Naval Research Laboratory, Washington, D.C. , 1979.

2. Oran, E.S., and Boris, J.P., Prog. Energy Combustion Science. 7:1-72 (1981).

3. Jones, W.W., and Boris, J.P., Comp. Chem. 5, 139-146 (1981).

4. Young, T.R., and Boris, J.P., J. Phys. Chem. 81, 2424-2427 (1977).

5. Young, T.R., CHEMEQ, A Subroutine for Solving Stiff Ordinary Differential Equations, N.R.L. Memorandum Report 4091, Naval Research Laboratory, Washington, D.C., 1980.

6. Williams, F.A., Combustion Theory, Addison Wesley, Reading, MA., 1965, p.2.

7. Picone, J.M., and Oran , E.S., Approximate Equations for Transport Coefficients of Multi-component Mixtures of Neutral Gases, N.R.L. Memorandum Report 4384, Naval Research Laboratory, Washington, D.C., 1980.

8. Chapman, S., and Cowling, T.G., The Mathematical Theory of Non-uniform Gases, Cambridge University Press, 1970.

9. Hirschfelder,J.O., Curtiss, C.F., and Bird, R. B., Molecular Theory of Gases and Liquids, John Wiley and Sons, Inc., New York, 1964.

10. Burks, T.L., and Oran, E.S., A Computational Study of the Chemical Kinetics of Hydrogen Combustion, N.R.L. Memorandum Report 4446, Naval Research Laboratory, Washington, D.C., 1980

11. Oran, E.S., Young, T.R., Boris, J.P., and Cohen, A., Weak and Strong Ignition-I, N.R.L. Memorandum Report 4664, Naval Research Laboratory, Washington, D.C., 1981. (Also to appear in Combust. Flame).

12. Kailasanath, K., Oran, E.S., and Boris, J.P., A Theoretical Study of the Ignition of Pre-Mixed Gases, 1982 (to appear in Combust. Flame.)

13. Stull, D.R., and Prophet, H., JANAF Thermochemical Tables, National Standard Reference Data Series, U.S. National Bureau of Standards, No. 37, 2nd Ed., Gaithersburg, Maryland, 1971.
14. Marrero, T.R., and Mason, E.A., J. Phys. Chem. Ref. Data, 1,3-118 (1972).
15. Svehla, R.A., Estimated Viscosities and Thermal Conductivities of Gases at High Temperatures, Technical Report No. R-132, NASA, Washington, D.C., 1962.
16. Dixon-Lewis, G., Combust. Flame, 36, 1-14 (1979).
17. Lewis, B., and von Elbe, G., Combustion, Flames and Explosions of Gases, Academic Press, New York, 1961, p. 326.
18. Warnatz, J. Ber. Bunsenges. Phys. Chem. 82, 643-649 (1978).

Mechanism of Flame Propagation in Hydrogen-Air and Methane-Air Systems

S.Fukutani and H.Jinno
Faculty of Engineering,
Kyoto University, Kyoto, Japan

Introduction

The propagation mode of laminar flames is expected to govern many properties in the flames. It has been well-known that the propagation mode is explained either by means of the "thermal theory" or of the "active-species-diffusion theory". Though these theories are satisfactory for roughly analyzing flame propagation, they aren't exact enough to discuss it in detail.

The thermal theory neglects the contribution of the chemical reactions in a preheat zone and ascribes the temperature increase in this zone only to thermal conduction. However at the boundary between a preheat zone and a reaction zone, the increment of thermal energy due to conduction becomes zero and, therefore, convectional heat flow and heat release through chemical reactions must be equal to each other as indicated by the equation of thermal energy conservation; some chemical reactions must take place at the boundary. Chemical reactions cannot start immediately when their environment satisfies certain conditions but are excited gradually. On the other hand, the flame propagation under the diffusion of active species depends explicitly upon chemical reactions. Chemical reactions should be eventually evaluated accurately for the detailed analyses of the both propagation modes. There is considerable difficulty in expressing analytically the rate of chemical reaction in flames without any experimental data (or data given by means of computer simulation). Moreover, actual flames propagate partly under thermal effects and partly under the diffusion of active species, that makes the problem more complicated.

In the early years of 1970's, Spalding *et al.* proposed a simulation model and this model has been applied to many flames [1-2]. Some investigators simulated H_2 and CH_4 flames and analyzed their structures [3-6].

The purpose of this investigation is to elucidate the factors determining the propagation mode of laminar premixed flames mainly from chemical viewpoint by simulating H_2-air and CH_4-air flames and by discussing their flame structures and propagation mechanisms in comparison with each other; these flames seem to be typical examples for the two propagation modes.

Mathematical Model

The following four assumptions were made for a mathematical model.

1. The flames under consideration are one-dimensional premixed flames.
2. The flow rate of gas mixture is small enough so that the pressure is constant throughout flames and is atmospheric pressure. Viscosity effects are negligibly small.
3. The heat loss from flames to their surroundings is neglected and then the temperature rises up to the adiabatic flame temperature at infinite distance. Radiative heat transfer is also neglected.
4. Hydrogen flames contain OH, H, O, HO_2, H_2O_2, H_2O, H_2, O_2, and N_2, and 21 elementary reactions (Table 1a). Methane flames comprise CH_4, CH_3, HCHO, CHO, CO, CO_2, OH, H, O, HO_2, H_2O, H_2, O_2, and N_2, and 33 reactions (Table 1b).

These assumptions simplify the governing equations like Eqs.1 to 4.

$$\partial\rho/\partial t + \partial/\partial x\cdot(\rho v) = 0 \tag{1}$$

$$c_p\rho\cdot\partial T/\partial t + c_p\rho v\cdot\partial T/\partial x = \partial/\partial x\cdot(\lambda\cdot\partial T/\partial x) + \rho\sum_i(D_i\cdot\partial h_i/\partial x\cdot\partial\omega_i/\partial x) - \sum_i(\phi_i h_i) \tag{2}$$

$$\rho\cdot\partial\omega_i/\partial t + \rho v\cdot\partial\omega_i/\partial x = \partial/\partial x\cdot(D_i\rho\cdot\partial\omega_i/\partial x) + \phi_i \tag{3}$$

$$p = \rho RT\sum_i(\omega_i/m_i) \tag{4}$$

Table 1a

Hydrogen-Oxygen Reaction Mechanism

$k = A \cdot T^n \cdot \exp(-E/T)$

No.	Reaction	A	n	E	Ref.
1	H_2 $+O_2$ $=OH$ $+OH$	2.50E+06	0.0	19600	7
2	H $+O_2$ $=OH$ $+O$	2.20E+08	0.0	8450	7
3	O $+H_2$ $=OH$ $+H$	1.80E+04	1.0	4480	7
4	OH $+OH$ $=O$ $+H_2O$	6.30E+06	0.0	550	7
5	OH $+H_2$ $=H$ $+H_2O$	2.20E+07	0.0	2590	7
6	H $+H$ $+M$ $=H_2$ $+M$	2.60E+06	-1.0	0	7
7	O $+O$ $+M$ $=O_2$ $+M$	1.90E+01	0.0	-900	7
8	H $+O$ $+M$ $=OH$ $+M$	3.60E+06	-1.0	0	8
9	OH $+H$ $+M$ $=H_2O$ $+M$	4.06E+10	-2.0	0	7
10	H $+O_2$ $+M$ $=HO_2$ $+M$	5.00E+03	0.0	-500	7
11	H $+HO_2$ $=H_2$ $+O_2$	2.50E+07	0.0	350	7
12	H $+HO_2$ $=OH$ $+OH$	2.50E+08	0.0	950	7
13	H $+HO_2$ $=O$ $+H_2O$	9.00E+05	0.5	2000	7
14	OH $+HO_2$ $=H_2O$ $+O_2$	5.00E+07	0.0	500	9
15	O $+HO_2$ $=OH$ $+O_2$	6.30E+07	0.0	350	10
16	HO_2 $+H_2$ $=H$ $+H_2O_2$	7.30E+05	0.0	9400	7
17	HO_2 $+HO_2$ $=H_2O_2$ $+O_2$	8.50E+06	0.0	500	7
18	OH $+H_2O_2$ $=HO_2$ $+H_2O$	1.00E+07	0.0	910	7
19	H $+H_2O_2$ $=OH$ $+H_2O$	2.20E+09	0.0	5900	7
20	O $+H_2O_2$ $=OH$ $+HO_2$	2.80E+07	0.0	3200	7
21	H_2O_2 $+M$ $=OH$ $+OH$ $+M$	1.20E+11	0.0	22900	7

k is expressed in m-mole-sec units.

Table 1b

Methane-Oxygen Reaction Mechanism

$k = A \cdot T^n \cdot \exp(-E/T)$

No.	Reaction	A	n	E	Ref.
1	CH_4 $+M$ $=CH_3$ $+H$ $+M$	8.00E+10	0.0	44500	11*
2	CH_4 $+OH$ $=CH_3$ $+H_2O$	2.80E+07	0.0	2500	11
3	CH_4 $+H$ $=CH_3$ $+H_2$	2.24E-02	3.0	4400	12
4	CH_4 $+O$ $=CH_3$ $+OH$	2.00E+07	0.0	4640	13
5	CH_3 $+O_2$ $=HCHO$ $+OH$	1.20E+05	0.0	5000	12
6	CH_3 $+O$ $=HCHO$ $+H$	1.00E+08	0.0	0	12
7	$HCHO$ $+M$ $=CO$ $+H_2$ $+M$	4.62E+10	0.0	17500	14*

Table 1b (continued)

No.	Reaction	A	n	E	Ref.
8	HCHO+OH =CHO +H_2O	5.40E+08	0.0	3170	12
9	HCHO+H =CHO +H_2	1.35E+07	0.0	1890	12
10	HCHO+O =CHO +OH	5.00E+07	0.0	2300	12
11	CHO +M =CO +H +M	2.00E+06	0.5	14400	11
12	CHO +O_2 =CO +HO_2	1.00E+08	0.0	3400	15
13	CHO +OH =CO +H_2O	1.00E+08	0.0	0	12
14	CHO +H =CO +H_2	2.00E+08	0.0	0	12
15	CHO +O =CO +OH	1.00E+08	0.0	0	12
16	CO +O_2 =CO_2 +O	3.00E+06	0.0	25000	16
17	CO +OH =CO_2 +H	5.60E+05	0.0	540	13
18	CO +O +M =CO_2 +M	2.36E+03	0.0	2184	10
19	CO +HO_2 =CO_2 +OH	1.50E+08	0.0	11900	8
20	H +O_2 =OH +O	2.20E+08	0.0	8450	7
21	O +H_2 =OH +H	1.80E+04	1.0	4480	7
22	OH +OH =O +H_2O	6.30E+06	0.0	550	7
23	OH +H_2 =H +H_2O	2.20E+07	0.0	2590	7
24	H +H +M =H_2 +M	2.60E+06	-1.0	0	7
25	O +O +M =O_2 +M	1.90E+01	0.0	-900	7
26	H +O +M =OH +M	3.60E+06	-1.0	0	8
27	OH +H +M =H_2O +M	4.06E+10	-2.0	0	7
28	H +O_2 +M =HO_2 +M	5.00E+03	0.0	-500	7
29	H +HO_2 =H_2 +O_2	2,50E+07	0.0	350	7
30	H +HO_2 =OH +OH	2.50E+08	0.0	950	7
31	H +HO_2 =O +H_2O	9.00E+05	0.5	2000	7
32	OH +HO_2 =H_2O +O_2	5.00E+07	0.0	500	9
33	O +HO_2 =OH +O_2	6.30E+07	0.0	350	10

k is expressed in m-mole-sec units.

* means that a frequency factor is modified.

where ρ is the density of gas mixture, v the flow rate, c_p the specific heat, T the temperature, p the pressure, λ the heat conductivity, h the enthalpy, ω the mass fraction, D the multicomponent diffusion coefficient, ϕ the production rate, m the molecular weight and the subscript i means the i-th species. The reaction schemes were selected to be applicable to various combustion conditions. The ploymerization, dehydrogenation and pyrolysis in CH_4 flames were not taken into consideration. The rate coefficients of the reverse reactions were obtained on the basis of the equilibrium constants evaluated from JANAF data[17]. The governing equations were solved numerically by means of a time-dependent method. The boundary conditions are that an unburnt gas is the mixture of fuel and air at room temperature with given flow velocity, and a burnt gas has the equilibrium composition at its adiabatic flame temperature. Binary diffusion coefficients and heat conductivities were estimated using approximate equations [18].

The H_2-air and the CH_4-air flames with a wide range of variation in the fuel contents were simulated and the obtained burning velocities were compared with the experimental values summarized by Andrews and Bradley [19] in order to verify the validity of the model. The H_2 content ranged from 10.0% to 70.0% in concentration, namely, from 3.771 to 0.1796 in air ratio. The CH_4 concentration was varied from 7.0% to 13.0%, that corresponds from 1.39 to 0.70 in air ratio. The obtained burning velocities agreed with the experimental velocities. The burning velocity calculated using a mathematical model should be determined as a result of the overall contributions of chemical and physical processes constituting the model. The good agreement permits to conclude that this model represents the combustion phenomena in actual H_2-air and CH_4-air premixed flames.

Results and Discussion

Figures 1a and 1b show the temperature profile and the concentration profiles of the unstable and the intermediate species in the stoichiometric H_2-air and CH_4-air flames. In the H_2 flame, H and O are present in larger quantities than in the CH_4 flame but the concentration of OH is almost the same in the both flames. Hydrogen atom in the H_2 flame is on the increase even at very low temperatures. Hydroxyl increases once at 400K, becomes almost constant and again increases at temperatures higher than 1000K. Hydroperoxyl and H_2O_2 have maxima in their concentrations at about 400K and 500K, though the values are much smaller. This suggests

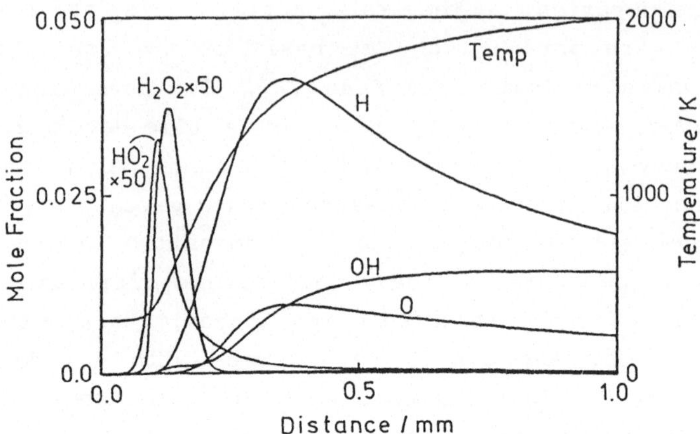

Fig.1a. Profiles of the temperature and the concentrations of unstable species in the stoichiometric H_2-air flame.

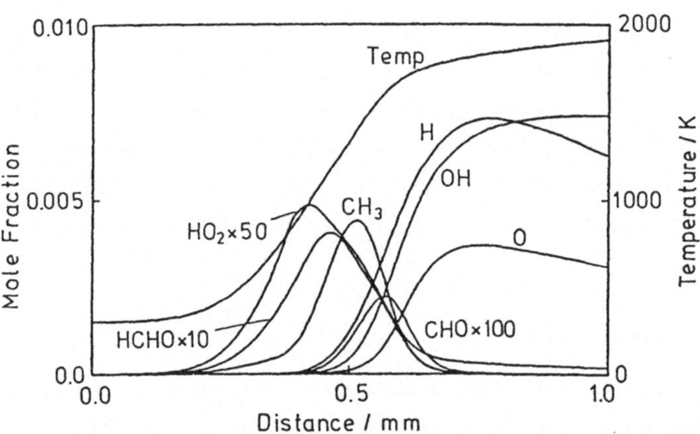

Fig.1b. Profiles of the temperature and the concentrations of unstable species in the stoichiometric CH_4-air flame.

that some chemical reactions already start in the low temperature region, and moreover the two-staged rise in OH concentration implies that the reactions in this flame consist of two different schemes separated at about 1000K. In the CH_4 flame, H, O and OH increase later comparing with the other species in the CH_4 flame or the same species in the H_2 flame.

The profiles of the rate of the each term in the equation of thermal energy conservation (Eq.(2)) in the stoichiometric H_2 and CH_4 flames are plotted in Figs.2a and 2b. The abscissas are expressed with the temperature for convenience of the direct comparison. In the CH_4 flame, the contribution of chemical reactions can be neglected at at 1200K or below and the inflow of thermal energy due to conduction is dominant. The H_2 flame, on the other hand, shows different appearances in these changes. The chemical reactions emit much heat even at an initial stage of the combustion. Its rate surpasses that of thermal conduction at 450K, becomes almost constant at about 700K and then decreases more gradually than that in the CH_4

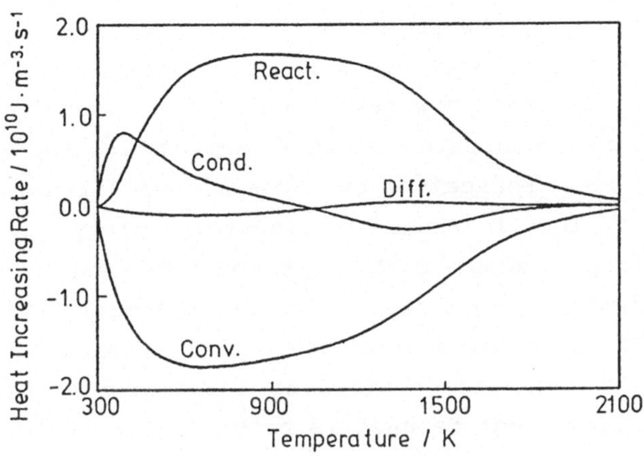

Fig.2a. Profiles of the temperature and the thermal energy balance in the stoichiometric H_2-air flame.

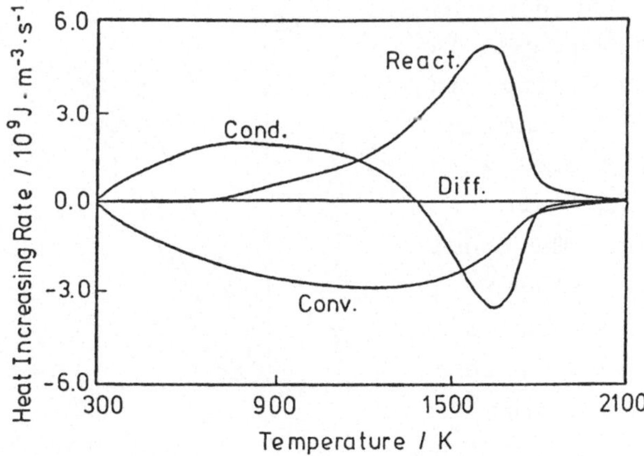

Fig.2b. Profiles of the temperature and the thermal energy balance in the stoichiometric CH_4-air flame.

flame. The rate of thermal conduction reaches its maximum at 400K. Thermal conduction doesn't contribute greatly to the heat transfer in this flame and, on the whole, the heat release due to reactions balances with the convectional heat flow. The intense heat release at the initial stage also suggests that chemical reactions take place in the low temperature region. The region where the heat increasing rate due to conduction is positive will be temporarily called a "preheat zone" by diverting the term used in the thermal theory, though it remains uncertain if this naming is appropriate also in the flames propagating by means of the diffusion of active species.

The increase in thermal energy in a preheat zone is brought about both by conduction and by heat release due to chemical reactions. In the thermally propagating flames, the effect of conduction must be dominant, but when flames propagate under the control of the diffusion of active species, the rate of heat release is probably larger than that of conduction. Then the index for the contribution of chemical reactions to the thermal energy increase in a preheat zone was expressed using the ratio of the heat release rate to the total heat increasing rate, and the mode of flame propagation was investigated. Figures 3a and 3b show the changes in this ratio in the H_2 and the CH_4 flames with various air ratios. The boundary between a preheat zone and a reaction zone is indicated as the ' position where the ratio is equal to one. In the H_2 flames with large burning velocities,

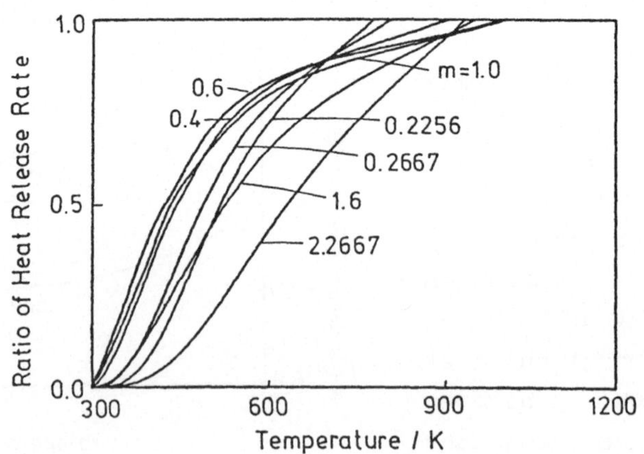

Fig.3a. Ratio of the heat release rate to the total heat increasing rate in H_2 flames.

such as the ones with air ratios of 0.4, 0.6 and 1.0, the intense heat release is observed even after a little temperature increase from those of the unburnt gases, the ratios become 0.5 at 430-450K and the boundaries are situated at rather high temperatures, 900-1000K (see also Fig.2a). Chemical reactions are expected to play a very important role especially in the propagation of these flames. An "ignition point" was defined as the position where the ratio of heat release rate becomes 0.5, because chemical reactions must occur with sufficient rates before gas mixture passes through the boundary and their contribution cannot be neglected beyond this point. In CH_4 flames the effect of heat release is negligibly small at temperatures lower than 700K. The temperature at the boundary is around 1400K with a few exceptions of the flames with large air ratios and that at the ignition point is around 1200K; the interval between these two points is only about 200° in temperature and chemical reactions don't emit much heat in preheat zones. The comparison of the changes in this ratio indicates clearly the difference of the role of chemical reactions in the propagation of H_2 and CH_4 flames.

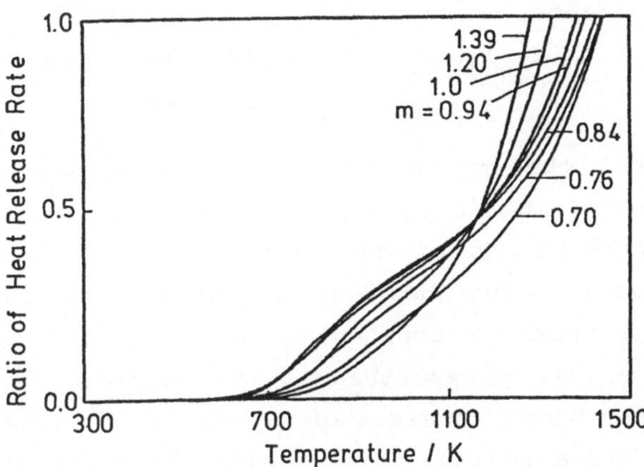

Fig,3b. Ratio of the heat release rate to the total heat increasing rate in CH_4 flames.

The sum of the rates of conduction and heat release must be equal to that of heat convection in a preheat zone of a stationary flame as evident by the definition of a preheat zone; in other words, gas mixture in a preheat zone is directly affected by the convection term, that is, by burning velocity, and then burning velocity is expected to depend much on the chemical and physical properties of the gas mixture in the preheat zone.

In H_2 flames, H_2 is oxidized exclusively by the reaction

$$OH + H_2 \rightarrow H + H_2O. \qquad (R1)$$

Then, the production and the consumption rates of OH in the stoichiometric H_2 flame are plotted in Fig.4, and the relative contributions of each reaction to its production and consumption are also shown in Figs.5a and 5b. Figures 4 and 5b indicate that the consumption of OH in the region of a large reaction rate is ascribed to (R1) alone. Its production, on the other hand, consists of two different scheme. In the low temperature region, OH is yielded by diffused H through the follow ing reaction sequence.

$$H + O_2 + M \rightarrow HO_2 + M \qquad (R2)$$

$$H + HO_2 \rightarrow OH + OH \qquad (R3)$$

These two reactions can oxidize H into OH readily even at lower temperatures near room temperature because of their negative and small activation energies, respectively. However, this reaction scheme cannot proliferate H by itself and the reaction rates depend on the quantity of diffused H. As temperature rises up, the competition reaction of (R2) becomes fast and the rates of these two reactions are equal to each other at around 1100K.

Then the chain reactions are activated and a large amount of H_2O and H are produced by reactions (R4), (R5) and (R1).

$$H + O_2 \rightarrow OH + O \qquad (R4)$$

$$O + H_2 \rightarrow OH + H \qquad (R5)$$

In the H_2 flames with large

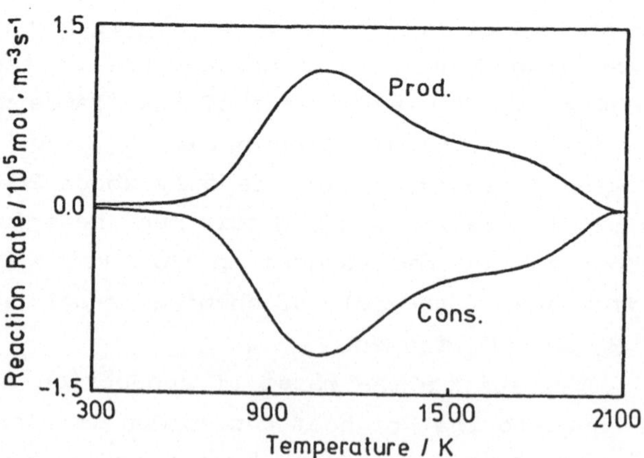

Fig.4. Production and consumption rates of OH in the stoichiometric H_2-air flame.

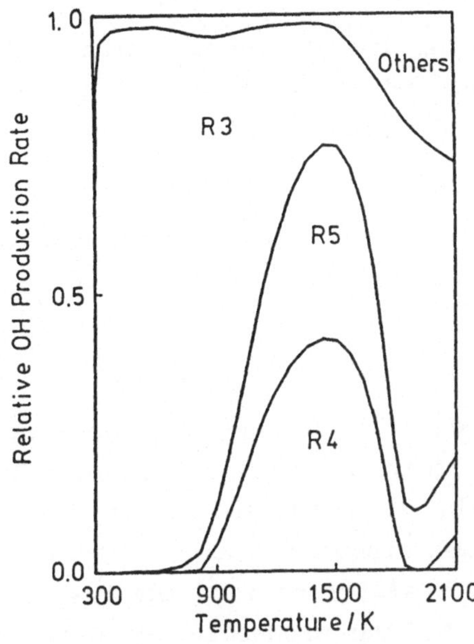

Fig.5a. Relative production rates of OH in the stoichiometric H_2-air flame.

$H+HO_2 \rightarrow OH+OH$ (R3)
$H+O_2 \rightarrow OH+O$ (R4)
$O+H_2 \rightarrow OH+H$ (R5)

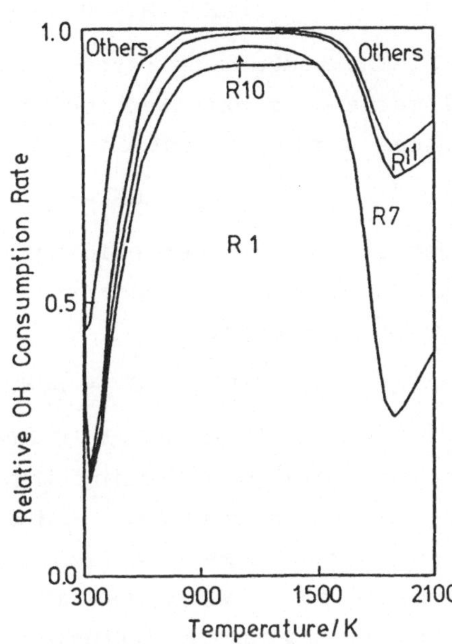

Fig.5b. Relative consumption rates of OH in the stoichiometric H_2-air flame.

$OH+H_2 \rightarrow H+H_2O$ (R1)
$OH+H+M \rightarrow H_2O+M$ (R7)
$OH+OH \rightarrow O+H_2O$ (R10)
$OH+HO_2 \rightarrow H_2O+O_2$ (R11)

burning velocity, those with air ratios of 0.4, 0.6 and 1.0, for example, the boundary between a preheat zone and a reaction zone agrees practically with the position where the rate of reaction (R4) becomes equal to that of (R2); gas mixture enters into its reaction zone after the excitation of the chain reactions. The heat release at low temperatures is ascribable to the reactions (R2), (R3) and (R6).

$$H + H + M \rightarrow H_2 + M \qquad (R6)$$

Hydrogen atom participates in all of these reactions and its diffusion governs the combustion reaction of H_2 in this region. The contribution of the reaction (R1) to the heat release becomes predominant around the boundary and then in the approach to the adiabatic flame temperature the recombination reactions (R7) and (R6) have large heat release rates.

$$OH + H + M \rightarrow H_2O + M \qquad (R7)$$

The combustion reactions of CH_4 are more complicated. All of methane is oxidized once to CH_4 and about 2/3 or 4/5 of CH_4 passes through reaction (R8) in the process of its oxidation.

$$CH_4 + OH \rightarrow CH_3 + H_2O \qquad (R8)$$

Most of CH_3 is then oxidized by reaction (R9), though some of it reacts with O_2 under the conditions of large air ratios and low temperatures.

$$CH_3 + O \rightarrow HCHO + H \qquad (R9)$$

After all both of OH and O are required for the combustion reactions of CH_4 including the further oxidation steps of HCHO or the other species. Hydroxyl is produced rather easily through the same reaction scheme discussed above under the presence of H, which can also be transferred into the low temperature region by diffusion. So the production of O holds the key to the combustion reactions of CH_4. The comparison of Figs.1b and 2b indicates that the temperature where the chemical reactions are activated agrees with that where O starts to increase. Figure 6 shows the rates of the production and the consumption of O in the stoichiometric CH_4 flame. These changes in the rate also support the conclusion obtained by the above comparison; that is, the rates of the reactions which O participates in increase at around 1200K, which is also the temperature of the ignition point of this flame. The effects of O on the initiation of the reactions were also inves-

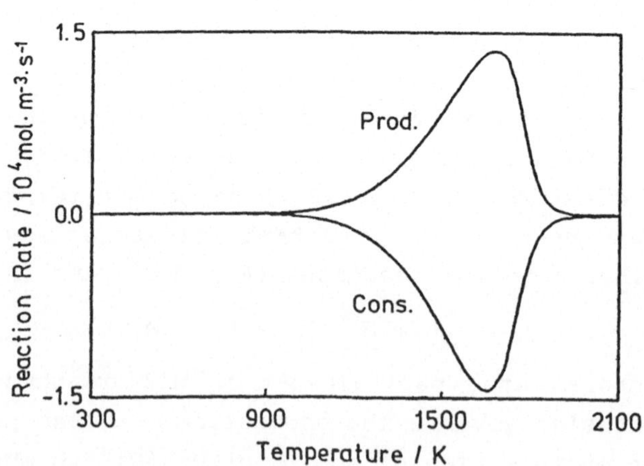

Fig.6. Production and consumption rates of O in the stoichiometric CH_4-air flame.

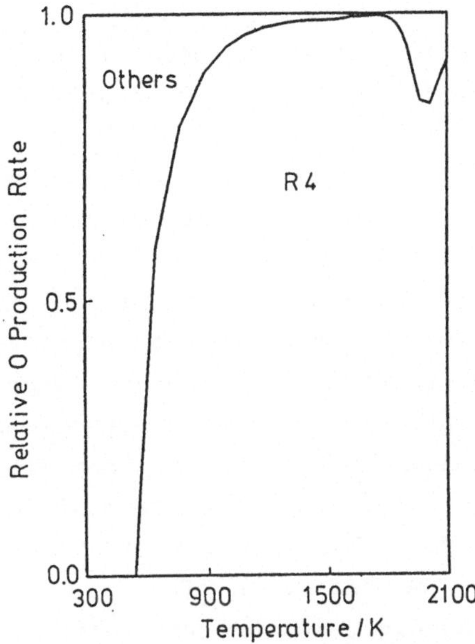

Fig.7a. Relative production rates of O in the stoichiometric CH_4-air flame.
$H+O_2 \rightarrow OH+O$ (R4)

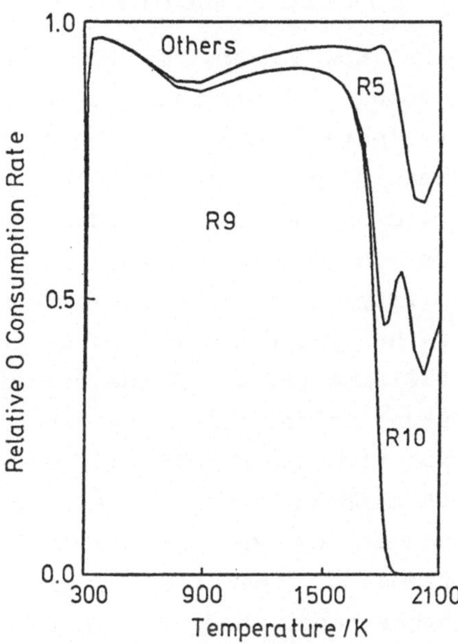

Fig.7b. Relative consumption rates of O in the stoichiometric CH_4-air flame.
$O+H_2 \rightarrow OH+H$ (R5)
$CH_3+O \rightarrow HCHO+H$ (R9)
$O+H_2O \rightarrow OH+OH$ (R10)

tigated by adding OH, H and O to the gas mixture with the same composition and the same temperature as those in the CH_4 flame and by comparing the change in the induction delay time. The addition of O can shorten that time to about a half of that obtained by the addition of the other two species. Then the production and the consumption mechanisms of O were analyzed. The relative contributions of each elementary reaction to its production and consumption are plotted in Figs.7a and 7b, respectively. According to these two figures, O is predominantly produced by reaction (R4) and consumed by reaction (R9) in the temperature region of 1200-1700K. Hydrogen atom is necessary for reaction (R4). However in temperatures lower than 1200K, H is oxidized through the competition reaction (R2). The ignition point, therefore, agrees with the temperature where these two reactions, (R2) and (R9), have the same rates.

Concluding Remarks

Now let us speculate on the criterion for determining the propagation mechanism of premixed flames. When O doesn't participate in any reaction steps with an important role, that is, when OH predominates alternatively as an oxidizer in the oxidation of fuel and in addition the activation energy of this oxidation is low enough, the combustion reaction has the possibility to take place even at low temperatures because of the rather rapid production of OH with diffused H through reactions (R2) and (R3). These reactions are exothermic and the temperature is raised as the reactions proceed. However, they cannot proliferate OH and may not lead to ignition by themselves in a short time. The increase in the reaction rate must rely heavily on that in the diffused H. The propagation of these flames is therefore governed by the diffusion of H. On the other hand, when O is an indispensable oxidizer, the initiation of the combustion reactions must wait until the temperature rises up to around 1200K, for O cannot be produced through reaction (R4) due to the precedence of the competition reaction (R2). Then the temperature must be raised by thermal conduction without any activation of the combustion reactions.

References

1. Patankar,S.V. and Spalding,D.B., Heat and Mass Transfer in Boundary Layer, 2nd ed., Intertext Books, London,1970
2. Stephenson,P.L. and Taylor,R.G., Combust. Flame 20, 231 (1973)
3. Dau,M.J.,Dixon-Lewis,G. and Thompson,K., Proc. R. Soc. Lond. A.330, 199 (1972)
   Dixon-Lewis,G., Goldworthy,F.A. and Greenberg,J.B., Proc. R. Soc. Lond. A.346, 261 (1975)
4. Smoot,L.D., Hecker,W.C. and Williams,G.A., Combust. Flame 26, 323 (1976)
5. Tsatsaronis,G., Combust. Flame 33, 217 (1978)
6. Warnatz,J., Eighteenth Symposium (International) on Combustion, p369, The Combustion Institute, Pittsburgh,1981
7. Baulch,D.L., Drysdale,D.D., Horne,D.G. and Lloyd,A.C.,

Evaluated Kinetic Data for High Temperature Reactions, vol.1, Butterworths, London, 1972

8. Jensen,D.E. and Jones,G.A., Combust. Flame 32, 1 (1978)
9. Heap,M.P., Tyson,T.J., Cichanowicz,J.E., Gershman,R. and Kau,C.J., Sixteenth Symposium (International) on Combustion, p535, The Combustion Institute, Pittsburgh, 1976
10. Baulch,D.L., Drysdale,D.D., Duxbury,J. and Grant,S., Evaluated Kinetic Data for High Temperature Reactions, vol.3, Butterworths, London, 1976
11. Bowman,C.T., Combust. Sci. Tech. 2, 161 (1970)
12. Bowman,C.T., Fifteenth Symposium (International) on Combustion, p.869, The Combustion Institute, Pittsburgh, 1974
13. Bowman,C.T., Fourteenth Symposium (International) on Combustion, p729, The Combustion Institute, Pittsburgh, 1973
14. Peeters,J. and Mahnen,G., Fourteenth Symposium (International) on Combustion, p.133, The combustion Institute, Pittsburgh, 1973
15. Burcat,A., Combust. Flame 28, 319 (1977)
16. Engleman,V.S., Bartok,W. and Longwell,J.P., Fourteenth Symposium (International) on Combustion, p.755, The Combustion Institute, Pittsburgh, 1973
17. Stull,D.R. and Prophet,H., JANAF Thermochemical Tables, 2nd ed., U.S. Dept. of Commerce, Washington, 1971
18. Perry,R.H. and Chilton,C.H., Chemical Engineers' Handbook, 5th ed., McGraw-Hill, New York, 1973
19. Andrews,G.E. and Bradley,D., Combust. Flame 20, 77 (1973)

Flames near rich flammability limits, with particular reference to the hydrogen - air and similar systems

By N.R. Carter, M.A. Cherian and G. Dixon-Lewis
Department of Fuel and Energy, The University, Leeds LS2 9JT, U.K.

1. Introduction

For mixtures of fuel and oxidant under defined initial conditions of temperature, pressure and environment, there exist well-defined composition limits of flammability between which premixed flame propagation will occur, but outside which a premixed flame will not propagate through the mixture indefinitely. Global reaction rate theories of the steady state propagation of infinite, adiabatic, planar flame fronts are unable straightforwardly to explain the existence of such limits, and indeed there is still no agreed explanation of why the limits occur. One approach has been to regard the limits as fundamental properties of the gas mixtures to which they refer; and following this, a number of attempts have been made [1-4] to relate the limits to some simple feature of the structure of the flames or the detailed flame chemistry. Some of these will be discussed below. An alternative view, supported by much experimental evidence showing the dependence of the limits on environmental conditions [5-8], is that the observed limits, which are measured for finite flames, are essentially quenching phenomena arising from some interaction of the flames with their environment. In support of such explanations, Spalding [9], Mayer [10], and Berlad and Yang [11] have shown by theoretical examination of non-adiabatic flames with heat losses that if, in a composition range, the reaction rate decreases with temperature more rapidly than the heat loss rate, then a composition limit of flammability is obtained, with a small but finite burning velocity at the limit. This prediction is in accord with experiment. Limit burning velocities are not zero, but are of the order of 5 cm s^{-1} or above [7,8].

Yet another possibility exists in the limit region [5]. This is that although the observed limits may be essentially quenching phenomena, a fundamental limit is nevertheless required by some internal condition in the flame; but that such a fundamental limit has never been obtained in practice due to masking by the observed limit. This last possibility can, of course, only be examined theoretically, and such an examination requires a detailed knowledge of the chemical mechanism in the flame. The detailed chemistry of hydrogen - oxygen - nitrogen flames is now rather firmly established [12], and the purpose of the present note is to draw attention to the fact that attempted solutions of the time-dependent equations for adiabatic, one-dimensional flames in hydrogen - oxygen - nitrogen mixtures begin to encounter stability problems, and to show oscillations, at compositions which correspond rather precisely with the observed flammability limits. The oscillatory behaviour is similar in nature to that shown by Shivasinsky [13] to occur in unstable flames supported by single global reactions having a strong temperature dependence of the reaction (high activation energy), when also, in his case, the Lewis number Le ($= \lambda/\rho D c_p$) > 1.

2. Computational procedures and input data

(a) General outline

Flame properties were computed by implicit solution of the time-dependent flame equations according to the approach initially developed by Patankar and Spalding [14] for the prediction of steady, two-dimensional, boundary layer flows; and later modified by Spalding, Stephenson and Taylor [15-17] to predict unsteady one-dimensional flame propagation. The latter modification has been refined by Tsatsaronis [18] to include detailed transport property calculations, using the equations given by Dixon-Lewis [19]. These are based on the extension of the Chapman-Enskog kinetic theory to polyatomic gases by Wang Chang, Uhlenbeck and de Boer [20], and the subsequent development by Mason, Monchick and coworkers [21-25].

(b) Reaction mechanism and rate parameters

The reaction mechanism and the expressions used for the forward rate coefficients are given in table 1. For consideration of the near-limit flames, the general hydrogen - oxygen flame reaction mechanism employed by Dixon-Lewis [12] has been extended to include reaction (x) which forms hydrogen peroxide, and reactions (xxi) to (xxvi) which remove it. The formation of hydrogen peroxide by reaction (x) may become important in near-limit flames where the concentrations of H, O and OH become very small. In this context it will also be recalled that the dissociation of hydrogen peroxide, reaction (xxvi), is an important step controlling the rate of the non-explosive reaction between hydrogen and oxygen in closed vessels [26]. It is therefore conceivable that it may also play an important part in the propagation of very slow flames.

The rate coefficients for the additional reactions were taken from Dixon-Lewis and Williams [26]. Equilibrium constants (required for the evaluation of the reverse rate coefficients) and other thermodynamic data

Table 1. Parameters of expressions for forward rate coefficients used in calculation. Rate coefficients are expressed as $k = A\,T^B \exp(-C/T)$ in cm mole s units.

	Reaction			A	B	C/K
(i)	$OH + H_2$	$H_2O + H$		1.17×10^{9}	1.3	1825
(ii)	$H + O_2$	$OH + O$		1.42×10^{14}	0	8250
(iii)	$O + H_2$	$OH + H$		1.8×10^{10}	1.0	4480
†(iv)	$H + O_2 + H_2$	$HO_2 + H_2$		1.03×10^{18}	-0.72	0
(vii)	$H + HO_2$	$OH + OH$		1.4×10^{14}	0	540
(viia)	$H + HO_2$	$O + H_2O$		1.0×10^{13}	0	540
(x)	$HO_2 + HO_2$	$H_2O_2 + O_2$		2.0×10^{12}	0	0
(xii)	$H + HO_2$	$H_2 + O_2$		1.25×10^{13}	0	0
(xiii)	$OH + HO_2$	$H_2O + O_2$		7.5×10^{12}	0	0
¶(xiv)	$O + HO_2$	$OH + O_2$	a	1.4×10^{13}	0	540
			b	1.25×10^{12}	0	0
(xv)	$H + H + H_2$	$H_2 + H_2$		9.2×10^{16}	-0.6	0
	$H + H + N_2$	$H_2 + N_2$		1.0×10^{18}	-1.0	0
	$H + H + O_2$	$H_2 + O_2$		1.0×10^{18}	-1.0	0
	$H + H + H_2O$	$H_2 + H_2O$		6.0×10^{19}	-1.25	0
(xvi)	$H + OH + M$	$H_2O + M$				
	$M = H_2, O_2, N_2$			1.6×10^{22}	-2.0	0
	$M = H_2O$			8.0×10^{22}	-2.0	0
(xvii)	$H + O + M$	$OH + M$				
	$M = H_2, O_2, N_2$			6.2×10^{16}	-0.6	0
	$M = H_2O$			3.1×10^{17}	-0.6	0
(xviii)	$OH + OH$	$O + H_2O$		5.75×10^{12}	0	390
(xxi)	$H + H_2O_2$	$HO_2 + H_2$		1.4×10^{12}	0	1800
(xxii)	$H + H_2O_2$	$OH + H_2O$		1.8×10^{14}	0	3900
(xxiii)	$OH + H_2O_2$	$H_2O + HO_2$		6.1×10^{12}	0	720

Table 1 (cont)

	Reaction		A	B	C/K
(xxiv)	$O + H_2O_2$	$OH + HO_2$	1.4×10^{13}	0	3200
(xxv)	$O + H_2O_2$	$H_2O + O_2$	1.4×10^{13}	0	3200
†(xxvi)	$H_2O_2 + H_2$	$OH + OH + H_2$	2.7×10^{17}	0	22900

† Chaperon efficiencies relative to H_2 = 1.0 are 0.44, 0.35 and 6.5 for N_2, O_2 and H_2O respectively [27].

¶ $k_{14} = k_{14a} + k_{14b}$

---

are based on JANAF Thermochemical Tables [27]. As with K_{15}, the expression for K_{10} was deduced by simple parametric fitting of the van't Hoff isochore to equilibriu- constants at 1500 and 2500K from the tabulation. The expressions for K_1, K_2 and K_3 are those due to Del Greco and Kaufman [28].

(c) Transport property calculations

The hydrogen - oxygen flame system involves nine chemical species when H_2O_2 is included in the mechanism. Multicomponent diffusion coefficients were calculated for the complete nine component system, according to the equations given by Dixon-Lewis [19]. In the calculation of the thermal conductivity and the multicomponent thermal diffusion coefficients, only the seven components H, O, OH, N_2, O_2, H_2 and H_2O were considered. Molecular interactions for the transport property calculations were represented by the Lennard-Jones (12:6) potential, with the force constants listed by Dixon-Lewis [12,29].

3. Results

Table 2 gives the results of attempted predictions of the properties of several hydrogen - air flames containing 70% hydrogen or more at atmospheric pressure, with T_u = 298 K. According to Burgoyne and Williams-Leir [30] the tube flammability limits for these conditions is at 73.8% hydrogen in a 4.8 cm diameter tube, and at 75.0% hydrogen in a 10 cm diameter tube. The omission of calculated flame properties in table 2 for unburnt compositions with $X_{H_2,u} > 0.751$ arises because initially no stable solutions of the equations could be obtained at these compositions, despite concerted attempts to do so by limiting the maximum allowable time interval of the computations to

10 μs, and at the same time allowing ample opportunity for a low burning velocity flame to expand to a steady state width (flame thickness being larger at lower burning velocities). The approach of the flame model to a steady state width is controlled in the computation by an "entrainment" formula which determines at each time step a flow of initial mixture into the cold end and a flow of combustion products away from the hot end of the grid. During the approach to the steady state these flows are not equal, and it is known [15-18] that instabilities may be produced if the flows are too sensitive to the instantaneous shapes of the developing flame profiles. An attempt to remove the observed instability by markedly reducing the sensitivity below the level normally used was also unsuccessful.

Table 2. Initially computed properties of adiabatic hydrogen - air flames near the rich flammability limit at atmospheric pressure, with T_u = 298 K.

Flame	A	B	C	D	E
$x_{H_2,u}$	0.70	0.75	0.751	0.752	0.755
T_b/K	1328.4	1165.1	1161.8	1158.5	1148.3
$S_{u,calc}/(cm\ s^{-1})$	81	34.9	33.7	-	-
† $10^4\ x_{H,max}$	123	41.2	-	-	-
At T/K	1199	1114	-	-	-
† Max. heat release rate / $(W\ cm^{-3})$	1620	413	-	-	-
At T/K	940	940	-	-	-
† Temperature at beginning of appreciable heat release / K	435	500	-	-	-

† These maxima and temperatures are only approximate, particularly for the start of the heat release zone.

Despite the lack of success of these attempts, it is still of course possible that the observed instability is a purely artificial one, and a more detailed examination of the flame profiles tended at first to confirm this suspicion. The net production of hydrogen atoms, essential for the propagation of the flame, was confined to the region surrounding just one nodal point in the grid. However, a first re-arrangement of the grid, which allowed the inclusion of three nodal points in the net production region, led only to an extension of the composition limit for normal solutions to

$X_{H_2,u}$ = 0.760. At richer compositions than this the oscillatory behaviour was again observed. A second re-arrangement provided for eight grid points in the H atom net production region. Again commencing solutions at $X_{H_2,u}$ = 0.752, and progressively and gradually enriching the flames in hydrogen in a smooth manner, the oscillatory behaviour commenced this time at $X_{H_2,u}$ = 0.759. The flame behaviour in the two sequences is summarized in table 3. It is virtually certain that the oscillations are genuine.

Table 3. Properties of solutions to the time-dependent adiabatic flame equations for hydrogen - air flames near the rich flammability limit.

$X_{H_2,u}$	$S_u/(\text{cm s}^{-1})$	T_b/K	$\rho_u S_u/(10^{-2}\ \text{kg m}^{-2}\ \text{s}^{-1})$	
	(a) With three grid points in H atom net production region			
0.759	28.1	1132	9.73	min
			9.78	max
0.760	27.4	1129	9.45	min
			9.49	max
0.761	-		7.0	min
			12.0	max
	(b)· With eight grid points in H atom net production region			
0.752	34.5	1156		
0.755	32.0	1146	11.20	min
			11.24	max
0.757	30.4	1139	10.60	min
			10.61	max
0.759	-		8.4	min
			12.4	max

Table 3 shows that the change in computational behaviour occurs very sharply indeed at an unburnt gas composition near to 76.0% hydrogen, that is, at a composition not far outside the experimentally observed limit of 75.0% hydrogen in a sufficiently wide tube. There is therefore at least a case that the flammability limit (which is measured for a flame with some radiative heat loss) is associated with the change in the adiabatic one-dimensional computational behaviour at a slightly richer composition. The "theoretical" limit flame temperature would become about 1139 K, and the burning velocity at the limit around 30 to 31 cm s^{-1}.

(b) Other hydrogen - oxygen - nitrogen flames

Both the supposed theoretical limit flame temperature and the theoretical limit burning velocity just quoted for the hydrogen - air system are considerably higher than those which may be observed when suitable hydrogen - oxygen - nitrogen flames are supported on an Egerton - Powling type of flat flame burner. These latter are flames which are heavily diluted with nitrogen as opposed to hydrogen. Dixon-Lewis and Williams [31] have been able to burn such flames with a theoretical flame temperature as low as 1039 K. For a flame having the initial mole fraction composition $X_{H_2,u} = 0.1889$, $X_{N_2,u} = 0.7677$ and $X_{O_2,u} = 0.0434$, with $T_u = 336$ K, the measured burning velocity was 5.9 cm s^{-1} (quoted on the basis of gases at 291 K/1 atm.). This flame was the richest that could be burned, for $X_{H_2,u}/X_{N_2,u} = 0.246$ and $T_u = 336$ K.

For these fixed values of $X_{H_2,u}/X_{N_2,u}$ and T_u, table 4 gives the results of attempted predictions of the burning velocities as the oxygen content of the initial mixture varies. Again there is evidence of instability at a position corresponding closely with the observed limit, though this still needs confirmation by means of computation with a re-arranged grid.

Table 4. Computed properties of adiabatic hydrogen - oxygen - nitrogen flames near the rich flammability limit at atmospheric pressure, with T_u = 336 K. In all flames $X_{H_2,u}/X_{N_2,u} = 0.246$.

Flame	F	G	H	I
$X_{H_2,u}$	0.1883	0.1889	0.1889	0.1891
$X_{N_2,u}$	0.7657	0.7677	0.7681	0.7689
$X_{O_2,u}$	0.0460	0.0434	0.0430	0.0420
T_b/K	1078.0	1038.7		
†$S_{u,calc}/(cm\ s^{-1})$	9.2	5.5	-	-

† In this table only, burning velocities are quoted on the basis of gases at 291 K/1 atm.

It is noteworthy in the present context that it has not been found possible experimentally to stabilize near-limit hydrogen - air flames as such on the Egerton - Powling flat flame burner. This observation is consistent with the apparent high theoretical limit burning velocity of these flames, which is above the maximum of about 15 cm s^{-1} that the burner can support.

4. Discussion

The burning velocities and properties of premixed hydrogen - oxygen supported flames can in general be satisfactorily explained theoretically by the use only of reactions (i) to (iv), (vii), and (xii) to (xviii) in table 1; that is, without the inclusion of reaction (x) and reactions (xxi) to (xxvi) of hydrogen peroxide. With the truncated mechanism it is of course essential that the overall reaction becomes effectively chain branching in some region of the flame, and the flames were found [(12)] to consist of four identifiable temperature regions:-

(i) A small preheat zone where a little heating occurs by thermal conduction alone,

(ii) The major heat release zone, where radicals produced in zone (iii), having diffused upstream, react with incoming gas by low activation steps. The gas in the early part of zone (ii) still also receives heat by thermal conduction,

(iii) A radical production zone where the temperature is high enough and the reactant concentrations are such that the system is effectively in a chain branching condition. In this zone the chain branching cycle moves towards partial equilibrium, and

(iv) The radical recombination region, where the system decays towards full equilibrium, subject to the partial equilibrium conditions eventually achieved near the end of zone (iii).

In all the hydrogen - air flames studied earlier by Dixon-Lewis [(12)], with compositions ranging from 15 to 70% hydrogen, effective radical production by chain branching commenced at about 900 K. It has also been found that in none of the flames does the inclusion of the additional reactions (x) and (xxi) to (xxvi) of hydrogen peroxide (table 1) either significantly alter the flame structure (except for the appearance of a small H_2O_2 concentration early in the flame), or increase the burning velocity by more than one per cent.

For the experimental limit flame composition containing 75% hydrogen, the burnt gas mole fractions are $X_{H_2,b} = 0.6807$, $X_{N_2,b} = 0.2085$ and $X_{H_2O,b} = 0.1108$. Since the contributions of reactions (xv), (xvi) and (xvii) to chain termination become negligible at low radical concentrations, the limit system becomes effectively branching when the rate of termination by the reactions of HO_2 near the burnt gas composition is balanced (or just exceeded) by the rate of chain branching there. For the quoted burnt gas composition the temperature at which this occurs is 940 K. The corresponding temperature for the limit flame of table 4 is 910 K. Both are much below the observed limit flame temperatures; and indeed, on the supposition of equality of the rates of chain branching and breaking at the hydrogen - air rich limit flame temperature, the limit mixture would contain approximately 81% hydrogen. It must be concluded that a net chain branching condition at the hot end of the reaction zone is a necessary, though not a sufficient prerequisite for the flame. It should also be noted that the reactions of hydrogen peroxide were included in the calculations of §3(a) and (b), and that again they make little difference to the result. The slow reaction between hydrogen and oxygen is clearly not sufficiently rapid to support a flame.

The question still remains as to what determines the observed limit for premixed flame propagation, and it is here that the onset of the oscillatory behaviour may be significant. The behaviour brings to mind the analtic finding of Sivashinsky [13], who showed that in the limit of large temperature dependence of its global controlling reaction rate (reaction A $\rightarrow$ B or similar), a flame becomes unstable when the Lewis number Le (= $\lambda/\rho D c_p$) is greater than unity. The behaviour is examined elsewhere in this volume. Near-limit hydrogen - oxygen supported flames certainly exhibit the large temperature dependence of the reaction rate. Because of the competition between chain branching and breaking, the rate of production of hydrogen atoms (a species of high diffusion coefficient) in the region immediately above 940 or 910 K (depending on the flame) is very highly temperature dependent.

References

1. Rosen, J.B.: J. Chem. Phys. 22, 733, 743 (1954); Sixth Symposium (Int) on Combustion, p.236. New York: Reinhold, 1956.
2. van Tiggelen, A.: Bull. Soc. Chim. Belg. 55, 202 (1947).
3. Weinberg, F.J. Proc. Roy. Soc. Lond. A 230, 331 (1955).
4. Linnett, J.W. and Simpson, C.J.S.M.: Sixth Symposium (Int) on Combustion, p.20. New York: Reinhold, 1956.
5. Powling, J.: Fuel, 28, 25 (1949).
6. Egerton, A.C. and Thabet, S.K.: Proc. Roy. Soc. Lond. A 211, 445 (1952).
7. Badami, G.N. and Egerton, A.C.: Proc. Roy. Soc. Lond. A 228, 297 (1955).
8. Dixon-Lewis, G. and Isles, G.L.: Seventh Symposium (Int) on Combustion, p.475. London: Butterworths, 1959.
9. Spalding, D.B.: Proc. Roy. Soc. Lond. A 240, 83 (1957).
10. Mayer, E.: Combustion and Flame, 1, 438 (1957).
11. Berlad, A.L. and Yang, C.H.: Combustion and Flame, 4, 325 (1960).
12. Dixon-Lewis, G.: Phil. Trans. Roy. Soc. Lond. A 292, 45 (1979).
13. Sivashinsky, G.I.: Int. J. Heat and Mass Transfer, 17, 1499 (1974).
14. Patankar, S.V. and Spalding, D.B.: Heat and mass transfer in boundary layers (2nd. ed). London: Intertext Books, 1970.
15. Spalding, D.B. and Stephenson, P.L.: Proc. Roy. Soc. Lond. A 324, 315 (1971).
16. Spalding, D.B., Stephenson, P.L., and Taylor, R.G.: Combustion and Flame, 17, 55 (1971).

17. Stephenson, P.L. and Taylor, R.G.: Combustion and Flame, 20, 231 (1973).
18. Tsatsaronis, G.: Combustion and Flame, 33, 217 (1978).
19. Dixon-Lewis, G.: Proc. Roy. Soc. Lond. A 307, 111 (1968).
20. Wang Chang, C.S., Uhlenbeck, G.E. and de Boer, J.: Studies in statistical mechanics (ed. J. de Boer and G.E. Uhlenbeck) vol 2. New York: John Wiley, 1964.
21. Monchick, L. and Mason, E.A.: J. Chem. Phys. 35, 1676 (1961).
22. Mason, E.A. and Monchick, L.: J. Chem. Phys. 36, 1622 (1962).
23. Monchick, L., Yun, K. and Mason, E.A.: J. Chem. Phys. 39, 654 (1963).
24. Monchick, L., Pereira, A.N.G. and Mason, E.A.: J. Chem. Phys. 42, 3241 (1965).
25. Monchick, L., Munn, R.J. and Mason, E.A.: J. Chem. Phys. 45, 3051 (1966).
26. Dixon-Lewis, G. and Williams, D.J.: Comprehensive chemical kinetics (ed. C.H. Bamford and C.F.H. Tipper) vol 17, p.1. Amsterdam: Elsevier, 1977.
27. JANAF Thermochemical Tables (2nd. ed.). National Bureau of Standards Publication NSRDS-NBS 37. Washington, D.C., 1971.
28. Del Greco, F.P. and Kaufman, F.: Ninth Symposium (Int) on Combustion, p.659. New York: Academic Press, 1963.
29. Dixon-Lewis, G.: Phil. Trans. Roy. Soc. Lond. A 303, 181 (1981).
30. Burgoyne, J.H. and Williams-Leir, G.: Proc. Roy. Soc. Lond. A 193, 525 (1948).
31. Dixon-Lewis, G. and Williams, A.: Combustion and Flame, 4, 383 (1960).

EXPERIMENTAL INVESTIGATION OF INTERNAL COMBUSTION OF VARIOUS FUELS

by

W. Müller

Institut für Kraft- und Arbeitsmaschinen
Universität Kaiserslautern, West-Germany

1. Engine Measurements

Investigations of the reaction processes during the combustion in engines and closed vessels have been the subject of many experimental and theoretical research works. First measurements of the combustion temperature in an Otto-engine were executed by G.M. Rassweiler and L. Withrow in 1935 using the line reversal method at the sodium line at 589 nm. They investigated the mechanism of knocking combustion.

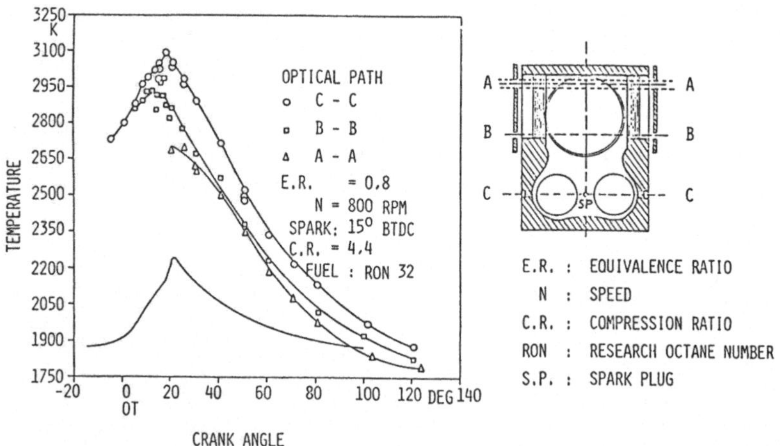

Figure 1: Temperature measurements during knocking combustion in an Otto-engine.
(G.M. Rassweiler, L. Withrow)

The investigations described in this paper were executed at the chair of combustion engines at the University of Kaiserslautern. The author thanks Professor Dr.-Ing. habil. H. May for the support of this work.

In Figure 1 the temperatures at different positions and the pressure are presented as an example. The spontaneous reaction of an unburnt portion of the mixture is indicated by a sharp pressure rise in front of the maximum.

A lot of experimental work in the field of engine combustion has been done since that time. In the following, results of investigations carried out at the University of Kaiserslautern are presented.

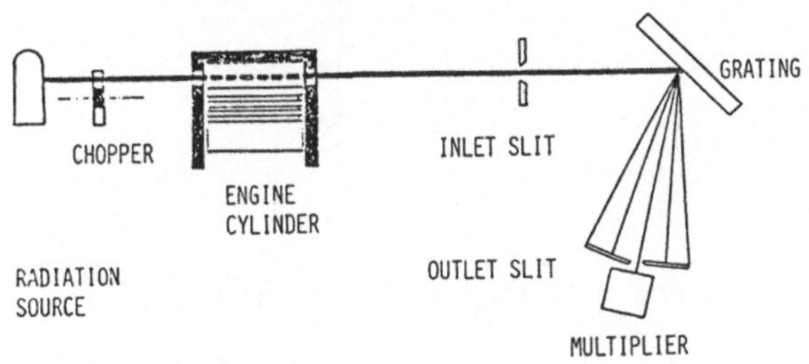

Figure 2: Optical arrangement for the execution of emission-absorption measurements at an Otto-engine.

Figure 2 shows the arrangement of the optical system used for the measurements of temperature and concentration in an Otto-engine. The results were calculated from emission and absorption data in the hydroxyl band system at 306.4 nm.

Using optical probes, as shown in Figure 3, it was possible to locate a small measurement region at different locations in the combustion chamber from the center near the spark plug to positions near the wall.

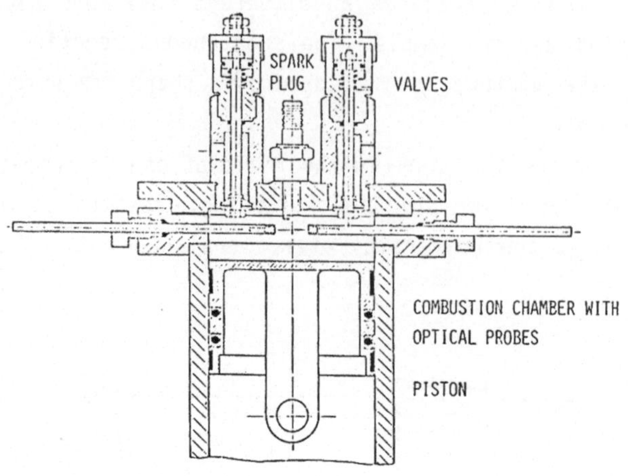

Figure 3: Test engine fitted with optical probes.

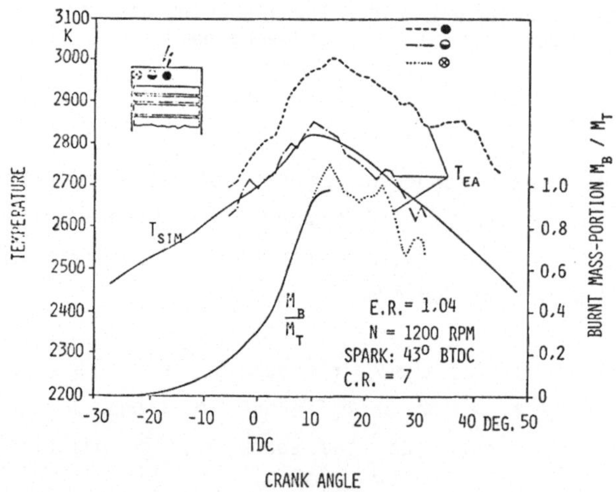

Figure 4: Emission-absorption temperatures at different engine combustion chamber positions.

Some results are presented in Figure 4. During the combustion the temperatures near the spark plug are about 300 K degrees above the values near the wall.

This difference is the result of different change of the thermal state during combustion and the following compression of the burnt gas mixture. The curve designated by T_{sim} is calculated by use of a simulation model on the basis of the measured pressure rise assuming a zone of combustion products with uniform temperature. The ratio of the burnt mass fraction m_b/m_t is also calculated from this model.

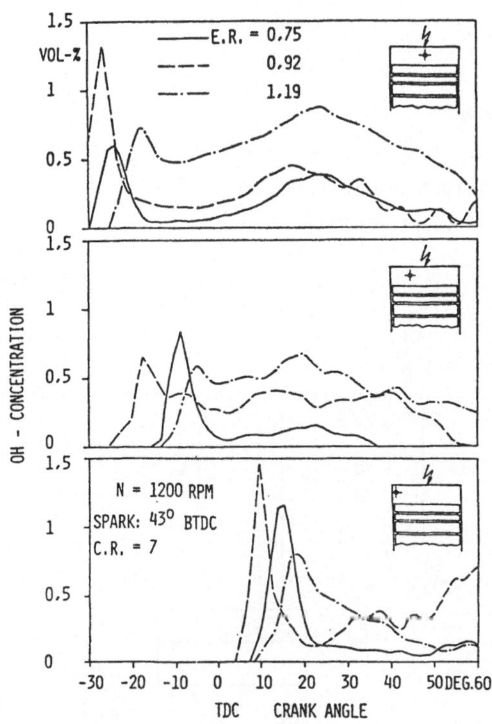

Figure 5: Course of the hydroxyl concentration at different locations in the combustion chamber for different equivalence ratios.

Figure 5 shows the course of the hydroxyl concentrations at the same locations in the combustion chamber. In the flame front region the OH-concentrations are rising rapidly up to a significant maximum which is far above the value for the chemical equilibrium state.

Behind this peak the concentration decreases rapidly in direction to the equilibrium values. These results mean that the reaction mechanism in the flame front region may be separated into a first region in which reactions predominantly are forming OH-radicals and a second one characterized by OH consumptioning reactions leading to the final products of combustion. The first region preliminary is responsible for the flame propagation due to high radical concentration gradients favoring the diffusion into the unburnt.

2. Constant Volume Bomb

Additional investigations were executed using a cylindrically combustion vessel of constant volume.

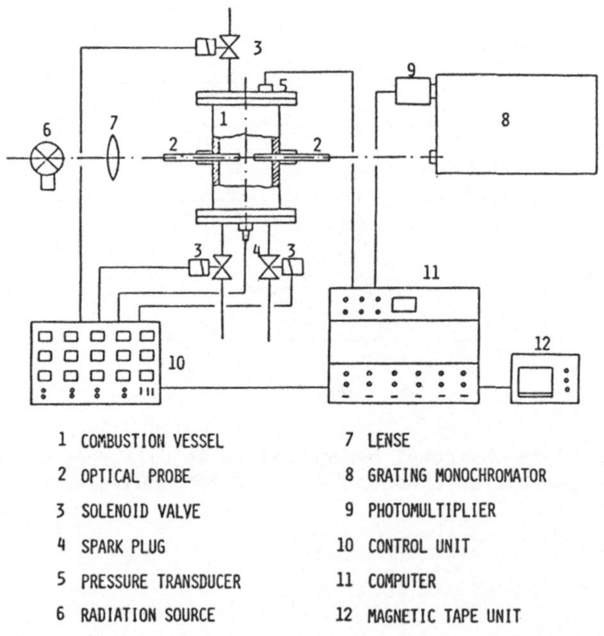

Figure 6: Arrangement of the test stand used for investigations of combustion in a closed vessel.

Figure 6 shows the arrangement of the test stand. Optical measuring points were located at different distances from the spark plug. The optical measurement procedure was analogous to the method used at the engine.
The absorption of the hydroxyl radicals at different locations in the vessel is represented in Figure 7 for the combustion of hydrogen air mixtures as a function of time.

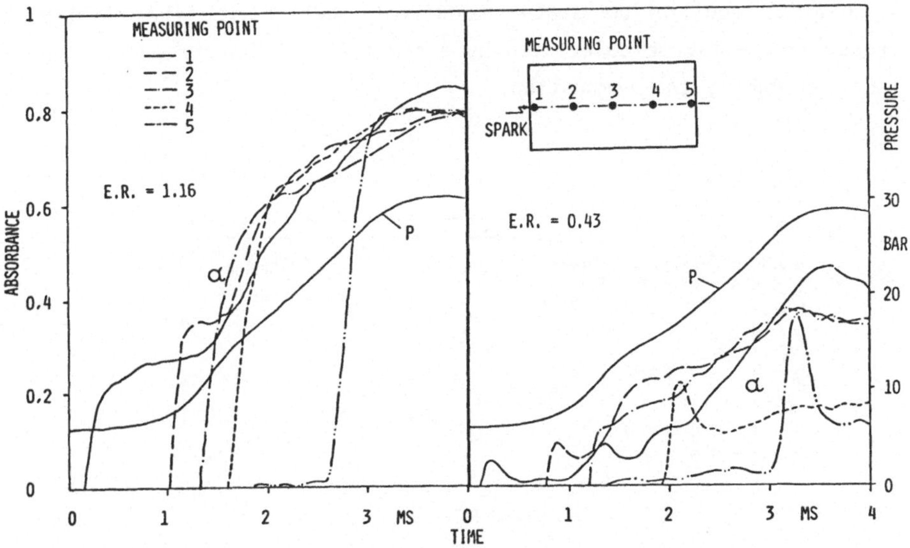

Figure 7: Course of the absorption of hydroxyl at 306.4 nm wavelength for different locations in the vessel and different equivalence ratios.

The arrival of the flame front is indicated by a sharp rise of the absorption at all points. But in the case of a lean mixture there are no absorption peaks as detected during the gasoline combustion in the engine. However, the peaks occur at hydrogen rich mixtures as shown in the right part of the figure. At location 5 the unburnt mixture is beginning to react before the flame front is arriving indicated by a small amount of OH produced at this point. This preflame reaction is caused by the high temperature level at this point due to the compression of the unburnt mixture.

Another interesting question is that about the amount of exothermical reactions in the flame front region. The total value of exothermical reactions in the vessel is indicated by the relative pressure rise which means the difference between the actual pressure and the initial pressue before ignition related to the difference of the maximum pressure and initial pressure.

This ratio is zero before ignition and equal to one after complete combustion. Assuming total energy conversion in the flame front with a chemical equilibrium state immediately behind it, the relative pressure-rise may be calculated by a simulation model for every flame location during the combustion.

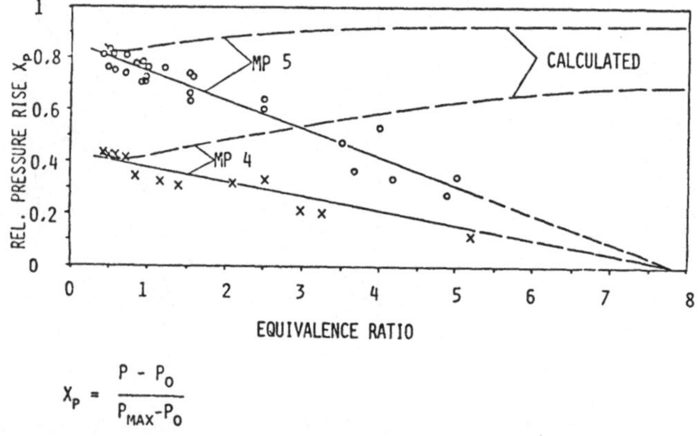

$$X_P = \frac{P - P_0}{P_{MAX} - P_0}$$

Figure 8: Relative pressure rise at the flame front arrival at two locations at the end of the vessel as a function of the equivalence ratio.

Figure 8 shows the results of such a calculation as dotted curves for the locations 4 and 5 at the end of the vessel as a function of the equivalence ratio (actual air/stoichiometric air).
Additionally some measured points are shown for both locations presenting the relative pressure at the flame front arrival indicated by the rise of the hydroxyl absorption signal.

As can be seen, there is an increasing difference between calculation and measurement with increasing equivalence ratio of the hydrogen air mixture. This result means that with increasing equivalence ratio a decreasing portion of exothermical reactions takes place in the flame front region. At lean mixtures the flame predominantly forms species permitting further reactions which need a relative long time until completion.

3. Calculation of Laminar Hydrogen Air Flames

The last figures show some results of laminar hydrogen-air flame calculations on the basis of a computer program developed by F. Schäfer. A small set of the following 7 reactions was regarded.

$OH + H_2 \rightleftharpoons H_2O + H$

$H + O_2 \rightleftharpoons OH + O$

$O + H_2 \rightleftharpoons OH + H$

$2O + M \rightleftharpoons O_2 + M$

$2H + M \rightleftharpoons H_2 + M$

$OH + H + M \rightleftharpoons H_2O + M$

$OH + OH \rightleftharpoons H_2O + O$

The calculation model uses an equidistant spaced grid of 200 locations and computation time intervals of 150 ns. Initial condition was an elevated temperature of 1500 K degrees at the first 3 elements.

Figure 9 shows the calculated curves of the temperature and the concentrations of atomic hydrogen, oxygen and the hydroxyl radical after a reaction time of 0.9 ms. The temperature increases uniformly beginning at the value of the unburnt mixture but the atomic hydrogen shows a very distinct maximum concentration in the flame front region which is far above the equilibrium value. Compared to the other species, the ascending flank of the hydrogen curve is displaced towards the unburnt mixture indicating the important influence of atomic hydrogen on the flame propagation due to its excellent diffusion property. This influence is confirmed by the results calculated for different equivalence ratios and presented in Figure 10.

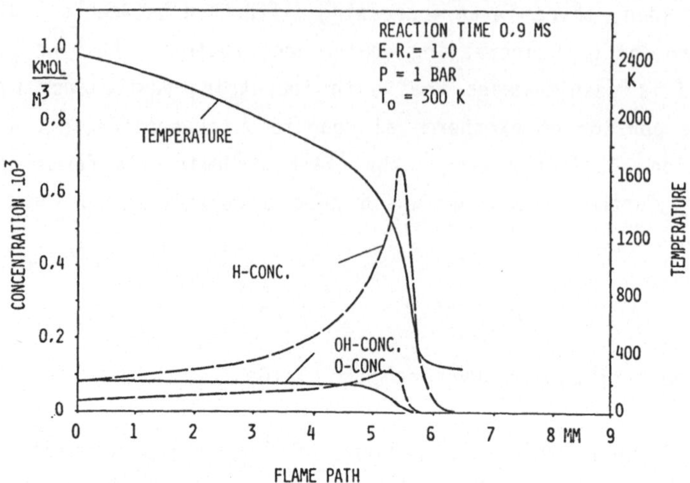

Figure 9: Calculated temperature, hydrogen-, oxygen- and hydroxyl concentration of a laminar flame front.

The peak values of the hydrogen concentrations vary as a function of the unburnt mixture composition arriving a maximum for an equivalence ratio of 0.6. With respect to this value increasing and decreasing equivalence ratios both are leading to decreasing hydrogen concentrations.

Furthermore, there is a direct relationship between the hydrogen peak values and the flame velocity indicated by the distance covered by the flame front after the time of 0.9 ms. For hydrogen air flames not the temperature level seems to be the most important influence on the flame propagation but the amount of atomic hydrogen production in the flame front.

The temperature curve of the lean mixture at an equivalence ratio of 2.0 shows a less slope than the others do. In agreement with the results presented in Figure 8, this indicates a decreasing amount of exothermical reactions in the prior flame region with increasing equivalence ratio for hydrogen lean mixtures.

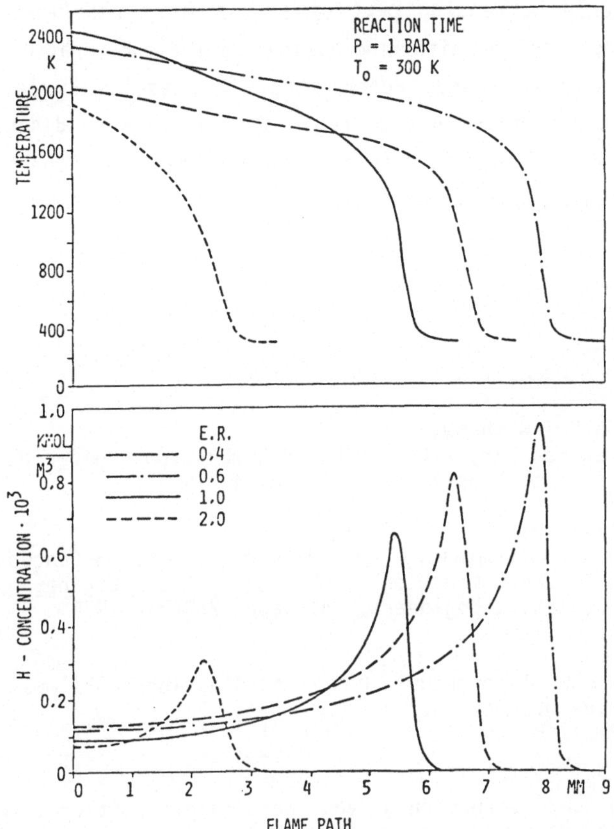

Figure 10: Calculated temperatures and hydrogen concentrations for laminar flames of different equivalence ratios.

4. Summary

In this paper the results of experimental investigations of the internal combustion in a test engine and a vessel of constant volume were represented. Some informations about the flame propagation and the exothermal flame front reactions were derived from spectroscopical measurements at the band system of the hydroxyl radical at 306.4 nm and from numerical simulation models.

A nearly linearly decreasing part of exothermical reactions in the flame front was found with increasing equivalence ratio in the hydrogen lean mixture range. Furthermore, the results of calculated laminar flame propagation in a hydrogen air system were discussed. A relation between the atomic hydrogen concentration and the velocity of the flame front could be detected.

References

1. Rassweiler, G.M.; Withrow, L.:
   Flame Temperatures Vary with Knock and Combustion Chamber Position.
   SAE Journal (Transactions), Vol. 36, No. 4, 1935

2. May, H.; Müller, W.:
   Spectroscopical Determination of Temperature- and OH-Concentration Distribution in the Combustion Chamber of an Otto- Engine.
   Society of Automotive Engineers, SAE Paper 780231, 1978

3. Müller, W.:
   Spektroskopische Temperatur- und Konzentrationsprofilmessung im Brennraum eines Ottomotors.
   MTZ 39 (1978), 7/8

4. Müller, W.:
   Untersuchung der Verbrennung von Wasserstoff-Luft-Gemischen in einer Verbrennungsbombe.
   Habilitationsschrift, Universität Kaiserslautern, 1981

5. Schäfer, F.; May, H.:
   Modell zur Berechnung einer laminaren Wasserstoff-Luft-Flamme.
   BWK 32, Nr. 12, 1980